Jesse Bering, PhD, is a frequent contributor to *Scientific American* and *Slate*. His writing has also appeared in *New York* magazine, The *Guardian*, and The *New Republic*, among other publications, and has been featured by NPR, Playboy Radio, the BBC, and more. The author of *The God Instinct*, Bering is the former director of the Institute of Cognition and Culture at Queen's University, Belfast, and began his career as a professor at the University of Arkansas. He lives in Ithaca, New York. His website is www.jessebering.com.

Why Is the Penis Shaped Like That?

And other reflections on being human

JESSE BERING, PhD

CORGI BOOKS

TRANSWORLD PUBLISHERS
61–63 Uxbridge Road, London W5 5SA
A Random House Group Company
www.transworldbooks.co.uk

WHY IS THE PENIS SHAPED LIKE THAT?
A CORGI BOOK: 9780552165792

First published in Great Britain
in 2012 by Bantam Press
an imprint of Transworld Publishers
Corgi edition published 2013

The majority of the essays in this book first appeared in different form and
as online columns, on *Scientific American*. The essays 'So Close, and Yet So
Far Away: The Contorted History of Autofellatio', 'Bite Me: The Natural
History of Cannibalism', 'Naughty by Nature: When Brain Damage Makes
People Very, Very Randy', and 'How the Brain Got Its Buttocks: Medieval
Mischief in Neuroanatomy' originally appeared, in different form, on *Slate*.

A CIP catalogue record for this book
is available from the British Library.

Addresses for Random House Group Ltd companies outside the UK
can be found at: www.randomhouse.co.uk
The Random House Group Ltd Reg. No. 954009

The Random House Group Limited supports The Forest Stewardship
Council® (FSC®), the leading international forest-certification organisation.
Our books carrying the FSC label are printed on FSC®-certified paper.
FSC is the only forest-certification scheme supported by the leading
environmental organisations, including Greenpeace. Our
paper-procurement policy can be found at
ww.randomhouse.co.uk/environment

Typeset in 10.5/13pt Sabon by Falcon Oast Graphic Art Ltd.
Printed and bound by CPI Group (UK) Ltd, Croydon, CR0 4YY.

2 4 6 8 10 9 7 5 3 1

For JCQ

Contents

An Invitation to Impropriety

For as long as I can remember, I've been sincerely curious, and vocally so, about certain "inappropriate" matters. My most earnest questions, I've noticed, tend to cause other people to back away from me very slowly. You might say I'm a little too analytical for my own good. "Isn't it unusual," I asked the absolutely horrified girl sitting next to me in my sixth-grade homeroom class one memorable day, "that my penis, when erect, is shaped more like a scimitar than a dagger? Certainly that must mean I'm deformed," I confided, whispering in her ear, "since obviously to penetrate a female like you properly, a penis must go straight into a vagina, not approach it from a forty-five-degree angle, as mine stands." Over time, I learned to bite my tongue. But a salacious mind, once stirred, seldom gets rest.

As I very slowly gained some much-needed social skills, I also found myself gravitating increasingly toward the world of science, a world in which nothing was sacred, no question too absurd or off-limits (at least for the sake of discussion, if not necessarily ethical fodder for the laboratory), and one in which I discovered other like-minded souls who didn't look at me as though I had three heads when I asked whether, say, people who prefer to be the recipient in anal intercourse might have a differently configured anal-genital internal anatomy than those who find it intensely unpleasant. I still don't know the answer to that question, by the way.

Speaking of which, I should probably also add (since it will become obvious enough by my disproportionate focus on male genitalia) that there was something very important to me that was denied full expression in my earlier years and that undoubtedly shaped my view of the world. I was gay. Very, very gay, in fact. I confirmed this incontrovertible truth through numerous experiments in my adolescence, including groping and kissing unwitting "girlfriends," who, in spite of their objective good looks and wonderful personalities, were as arousing to me as a perfumed slab of ham with sparkling white teeth. This wasn't just the virginal jitters, I can assure you, but girls seemed to make my penis positively catatonic, while even from afar boys made it stand at that oddly forty-five-degree angle I mentioned before.

So, let me start by offering a full disclosure: my perspective is that of a godless, gay, psychological scientist with a penchant for far-flung evolutionary theories. Still, although I certainly don't try to hide my own personal convictions, I'm an impolitic person. All I ask is that you try to suspend judgment until after you've read at least a handful of essays. Just lean back, unbutton your pants, and, by all means, get comfortable with yourself. Maybe relax with a glass of Chardonnay. And *think*. I hope to make that last part easy for you. I want you to enjoy learning about your wildly ejaculating penises, your dribbling vulvae, and your own fears, biases, fetishes, and desires. Despite our differences, and there are certainly many in this world, there's one thing we all have in common: we're human.

I'm not interested in sensationalism for its own sake, but

many of the questions that appeal to me most are, by definition, rather sensational. If you look at them closely enough, however, you'll notice how often the most titillating topics are uniquely able to raise deeper philosophical questions and to bring much more substantial issues to the surface. For instance, in reading about zoophiles, you may find yourself, as I did, questioning your own knee-jerk moralistic sexual repulsions; a look at the evolution of pubic hair or acne unexpectedly reveals our close genetic relationship with other apes; masturbation fantasies reveal what makes us unique in the animal kingdom; and foot fetishists expose how our adult turn-ons are permanently calibrated by often-innocent childhood experiences.

I do try to be a good scientist first and foremost, whether I'm investigating female ejaculation, six-month-old infants unexpectedly sprouting pubic hair, or the psychology of women curiously entranced by gay men. Since many of these essays were published originally, in some form, in my columns at *Scientific American* and *Slate* magazines, and therefore survey only the most interesting dimensions of any given topic, I'm certainly not able to cover every aspect and contrary viewpoint surrounding every issue. I encourage you, however, to read further about the subjects that leave you wanting more, and for that purpose I've included endnotes to help get you going.

So, please, join me in impropriety. Let's not subscribe to the some-things-are-better-left-unsaid school of life. How very boring that must be. I invite you to follow along with me on a journey of scientific discovery. Feel free to dip in and out in your reading or read the essays out of sequence. Each stands alone. But do watch your step: it's a slippery one. And note that although the mood is for the most part light, it won't be all fun and games. Some of the essays I've included in this anthology are actually rather sobering—

including a *really* close look at the mind-set of a suicidal person. I wrote that particular piece in response to the alarming rash of gay teen suicides in recent years. That was an article that resonated, and unfortunately so, with many readers, some of whom courageously shared their personal stories with me after stumbling upon it.

There are eight sections in this volume, each one representing a general theme or subject area and sampling the astounding oddities of simply being human. The first of these sections, "Darwinizing What Dangles," includes everything you didn't know you always wanted to know about male reproductive anatomy. In part II ("Bountiful Bodies"), we'll examine how we may be designed by Mother Nature to consume each other's flesh, why we're the only ape that suffers from acne, and many other little-known things about seemingly banal body parts. Next, in part III ("Minds in the Gutter"), we'll explore some very dirty brain science, pushing our common sense into a few uncomfortable corners in the process. This prepares us for part IV ("Strange Bedfellows"), where we'll take a critical, nonjudgmental look at some of the more intriguing sexual paraphilias, fetishes, and conditions, exploring their developmental origins, theories, and debates regarding clinical diagnoses. If you think having sex with animals is inherently wrong, or that sexuality starts in adolescence with the first flush of hormones, you may come away from this section with an unexpected change of mind.

In "Ladies' Night" (part V), we'll turn our attention specifically to the minds and bodies of women. Just note that I'm a gay man looking into and at these minds and bodies, so my take is a bit different from most. Speaking of which—and I'm not sure what Nietzsche would have to say about the content of the following section—in part VI ("The Gayer Science: There's Something Queer Here"), we'll then focus on some of the latest and most provocative studies on homosexuality. In part VII ("For the Bible Tells Me So"), we'll examine how religion stems from our evolved psychology and how our standard burial practices aren't doing ourselves or the planet any favors. And finally, in the last section of the book, "Into the Deep: Existential Lab Work," we'll dig into some weighty, soul-wrenching questions about suicide, the meaning of life, and the evolution of joy and happiness.

Excited? I hope so. And what better place for us to start than by asking why in the world testicles hang like that—and why does it hurt so *much* to get kicked there?

PART I

Darwinizing What Dangles

How Are They Hanging?
This Is Why They Are

A few years ago, the evolutionary psychologist Gordon Gallup, whom we'll meet again later in this section, along with his colleagues Mary Finn and Becky Sammis, set out to explain the natural origins of the only human male body part arguably less attractive than the penis—the testicles. In many respects, their so-called activation hypothesis elaborates on what many of us already know about descended scrotal testicles: they serve as a sort of cold storage and production unit for sperm, which keep best at a temperature slightly lower than the norm for the rest of our bodies. But the activation hypothesis goes much further than this fun fact.

It turns out that human testicles display some rather elaborate yet subtle temperature-regulating features that have gone largely unnoticed by doctors, researchers, and laymen alike. The main tenet of the activation hypothesis is that the heat of a woman's vagina radically jump-starts sperm that have been hibernating in the cool, airy scrotal sac. This heat aids conception. Yet it explains many other things too, including why one testicle is usually slightly lower than the other, why the skin of the scrotum some-times becomes rugose (prune-like and as wrinkled as an elephant's elbow), why the testicles retract during sexual arousal, and even why testicular injuries—compared with other types of bodily assaults—are so excruciatingly painful.

To help us all get on the same page, consider an alternate reality, one in which ovaries, like testicles, descend during embryological development and emerge outside the female body cavity in a thin, unprotected sac. After you've wiped that image from your mind's eye, note that the dangling gonads of many male animals (including humans) are no less puzzling. After all, why in all of evolution would nature have designed a body part with such obviously enormous reproductive importance to hang outside the body, so defenseless and vulnerable? We tend to become accustomed to our body parts, and it often fails to occur to us to even ask why they are the way they are. Some of the biggest evolutionary mysteries are also the most mundane aspects of our lives.

So the first big question is why so many mammalian species evolved hanging scrotal testicles to begin with. The male gonads in some phylogenetic lineages went in completely different directions, evolutionarily speaking. For example, modern elephant testicles are deeply embedded in the body cavity (a trait referred to as testicond), whereas other mammals, such as seals, have descended testicles but are without scrota, with the gonads simply being subcutaneous.

Gallup and his colleagues jog through several possible theories of our species' testicular evolution by descent. One of the more fanciful accounts—and one ultimately discarded by the researchers—is that scrotal testicles evolved in the same spirit as peacock feathers. That is to say, given the enormous disadvantage of having your entire genetic potential contained in a thin satchel of unprotected, delicate flesh and swinging several milli-meters away from the rest of your body, perhaps scrotal testicles evolved as a sort of ornamental display communi-cating the genetic quality of the male. In evolutionary biology, this type of adaptationist account appeals to the handicapping principle. The theoretical gist of the handi-capping principle is that if the organism can thrive and survive while still being hobbled by a costly, maladaptive trait such as elaborate, cumbersome plumage or (in this case) vulnerably drooping gonads, then it must have some high-quality genes and be a valuable mate.

But the handicapping hypothesis doesn't quite fit the case of descended scrotal testicles, explain the authors, because if it were true, then we would expect to see these body parts becoming increasingly elaborate and dangly over the course of evolution, not to mention that women should display a preference for males toting around the most ostentatious scrotal baggage. "With the possible exception of colored scrota among a few species of primates," writes Gallup, "there is little evidence that this has been the case." I'm not aware of any studies on intraspecies individual variation in scrotal design, but I'm nonetheless willing to speculate that most human males have rather bland, run-of-the-mill scrota. Anything deviat-ing from this—particularly a set of unusually pendulous testicles suspended in knee-length scrota—is probably more likely to have a woman dry heaving, screaming, or staring in confusion rather than serving as an aphrodisiac.

Again, a more likely explanation for scrotal descent, and one that has been around for some time, is that sperm production and storage are maximized at cooler temper-atures. "Not only is the skin of the scrotal sac thin to promote heat dissipation," the authors write, "the arteries that supply blood to the scrotum are positioned adjacent to the veins taking blood away from the scrotum and function as an additional cooling/heating exchange mechanism. As a consequence of these adaptations average scrotal temperatures in humans are typically 2.5 to 3 degrees Celsius lower than body temperature (37 degrees Celsius), and spermatogenesis is most efficient at 34 degrees Celsius."

Sperm are extraordinarily sensitive to even minor fluctuations in climate. When the ambient temperature rises to body levels, there is a momentary increase in sperm motility (they become more lively), but only for a period of time before fizzing out. To be more exact, sperm thrive at body temperature for fifty minutes to four hours, the length of time it takes for them to journey through the female reproductive tract and fertilize the egg. But once the spermatic atmosphere rises much above 37 degrees Celsius, the chances for a successful insemination consequently plummet—any viable sperm become the equivalent of burned toast. So in other words, except during sex, when it's adaptive for sperm to be hyperactive, sperm are stored and produced most efficiently in the cool, breezy surroundings of the relaxed scrotal sac. One doesn't want his scrotum to be too cold, however, since nature has calibrated these temperature points at precisely defined optimal levels.

Fortunately, human scrota don't just hang there holding our testicles and brewing our sperm; they also "actively" employ some interesting thermoregulatory tactics to protect and promote males' genetic interests. I place "actively" in scare quotes, of course, because, although it would be rather odd to ascribe consciousness to human scrota, testicles do respond unintentionally to the reflexive actions of the cremasteric muscle. This muscle serves to retract the testicles so they are drawn up closer to the body when it gets too cold—just think cold shower—and also to relax them when it gets too hot. This up-and-down action happens on a moment-to-moment basis; thus male bodies continually optimize the gonadal climate for spermatogenesis and sperm storage. It's also why it's generally inadvisable for men to wear tight-fitting jeans or especially snug "tighty whities"; under these restrictive conditions the testicles are shoved up against the body and artificially warmed so that the cremasteric muscle cannot do its job properly. Another reason not to wear these things is that it's no longer 1988.

Now, I know what you're thinking. "But, Dr. Bering, how do you account for the fact that testicles are rarely perfectly symmetrical in their positioning within the same scrotum?" In fact, the temperature-regulating function governed by the cremasteric muscle can account even for the most lopsided, one-testicle-above-the-other, waffling asymmetries in testes positioning. According to a 2009 report in *Medical Hypotheses* by the anatomist Stany Lobo and his colleagues, each testicle continuously migrates in its own orbit as a way of maximizing the available scrotal surface area that is subjected to heat dissipation and cooling. Like ambient heat generated by individual solar panels, when it comes to spermatic temperatures, the whole is greater than the sum of its parts. With a keen enough eye, presumably one could master the art of "reading" testicle alignment, using the scrotum as a makeshift room thermometer. But that's just me speculating.

From an evolutionary perspective, the design of male genitalia makes sense only to the extent that it adaptively complements the female anatomy, which, I realize, I should really go into more (but there are only so many hours in a day). By contrast to males, unless a woman is engaging in strenuous exercise, the female reproductive tract is maintained continuously at standard body temperature. This is the crux of Gallup's activation hypothesis: the rise in temperature surrounding sperm as occasioned by ejaculation into the vagina "activates" sperm, temporarily making them frenetic and therefore enabling them to acquire the necessary oomph to penetrate the cervix and reach the fallopian tubes. "In our view," write the authors, "descended scrotal testicles evolved to both capitalize on this copulation/insemination contingent temperature enhancement and function to prevent premature activation of sperm by keeping testicular temperatures below the critical value set by body temperatures."

One of the things you may have noticed in your own genitalia or those of someone you're especially close to is that in contrast to the slackened scrotal skin accompanying flaccid, nonaroused states, penile erections are usually accompanied by a telltale retraction of the testicles closer to the body. (This is the sort of thing easiest to demonstrate using visual illustrations, and a quick Google image search should provide ample examples. Just choose your own search terms and disable "safe search"—though if you're out in public right now, you may want to save this as homework for later.) According to Gallup and his co-authors, this is another smart scrotal adaptation. Not only does the cremasteric reflex serve to raise testicular temperature, thus mobilizing sperm for pending ejaculation into the vagina, but (added bonus) it also offers protection against possible damage to too-loose testicles resulting from vigorous thrusting during intercourse.

There are many other ancillary hypotheses connected to the activation hypothesis as well. For example, the authors ponder whether humans' well-documented preference—

and one rather unique in the animal kingdom—for night-time sex can be at least partially explained by temperature-sensitive testicles. Although the authors note the many additional benefits of nocturnal copulation (such as accommodating clandestine sex or minimizing the threat of predation), this preference may also reflect a circadian adaptation related to descended scrota. Given that our species evolved originally in equatorial regions where daytime temperatures often soared above body temperature, optimal testicular adjustments would be difficult to maintain in such excessive heat. In contrast, ambient temperatures during the evening and at night fall below body temperature, returning to ideal thermo-regulatory conditions for the testes. Additionally, after nighttime sex the female partner is likely to sleep, thus remaining in a stationary, often supine position that also maximizes the odds of fertilization.

Although the activation hypothesis helps us to better understand the functional, if quirky, architecture of the human male gonads, it may still seem odd to you that nature would have invested so heavily in such a precipitously placed genetic bank. After all, we're still left with the curious fact that these precious gametes are literally hanging in the balance in a completely unprotected vessel. Gallup and his coauthors weigh in on this, too:

Any account of descended scrotal testicles must also address the enormous potential costs of having the testicles located outside the body cavity where they are left virtually unprotected and especially vulnerable to insult and damage. To be consistent with evolutionary theory the potential costs of scrotal testicles would have to be offset not only by compensating benefits (e.g., sperm activation upon insemination), but one would

also expect to find corresponding adaptations that
function to minimize or negate these costs.

Enter pain. Not just any pain, but the unusually acute,
excruciating pain accompanying testicular injury. Most
males have some horrific stories to tell on this score—
whether it be a soccer ball to the groin or the flailing foot
of a sibling—but all of us men have something in common:
we've all become extraordinarily hypervigilant against
threats to the welfare of our scrotal testicles. According to
the authors, the fact that males are so squeamish and
sensitive regarding this particular body part can again be
understood in the context of evolutionary biology. If
you're male, the reason you probably wince more when
you hear the word "squash" or "rupture" paired with
"testicle" than you do with, say, "arm" or "nose" is that
testicles are disproportionately more vital to your reproductive success than these other body parts. I, for one, had
to pause to cover myself even typing those words together.
It's not that those other body parts aren't adaptively
important or that it doesn't hurt when they're injured.
Rather, it's a question of the *degree* of pain. Variation in
pain sensitivity across different bodily regions, according
to this view, reflects the vulnerability and importance that
different adaptations play in your reproductive success.
Many children have been born of broken-nosed men, but
not a single one has ever been sired by a man with two
irreparably damaged testicles. The point is that male
ancestors who learned to protect their gonads would have
left more descendants, and pain is a pretty good motivator
for promoting preemptive defensive action. Or to think
about it another way: any male in the ancestral past who
was oblivious to or freakishly enjoyed testicular insult
would have been quickly weeded out of the gene pool.

The wonders of the cremasteric muscle don't end here. It also flexes in response to threatening stimuli, in effect pulling the testicles up closer to the body and out of harm's way. In fact, the authors point out, Japanese physicians have been known to apply a pinprick to the inner thigh of male patients as a surgical prep: if the patient displays no cremasteric reflex, the spinal anesthesia has kicked in, and he's ready to go under the knife. Other evidence suggests that fear and the threat of danger trigger the cremasteric reflex. There are a number of ways to test this at home, if you're so inclined. Just make sure the owner of the fearfully reflexive testicles knows what you're up to before frightening him.

So, there you have it—an evolutionarily informed account of descended scrotal testicles in humans. Is the whole thing nuts? Don't leave me hanging, folks. Ball's in your court.

So Close, and Yet So Far Away:
The Contorted History of Autofellatio

Long before I knew very much about anything regarding sex, I did what many young males do, which of course is to place an empty paper-towel roll over my penis and suck hopefully upon the cardboard end. Okay, perhaps not everyone does this; I was a little confused about the suction principle. And now I'm a bit embarrassed by the story, although it's been a full year since the event and I'm much better informed on the subject of fellatio today. Oh, settle down, I'm only joking.

Well, *kind of*. I did actually attempt this feat, but I was twelve or thirteen at the time, which, to give you a clearer sense of my unimpressive carnal knowledge at that age, is also around the time that I submitted to my older sister with great confidence that a blow job involves using one's lips to blow a cool breeze upon another's anus.

So to avoid similar confusion, let us define our terms clearly. Autofellatio, the subject at hand—or rather, not at hand at all—is the act of taking one's genitals in one's mouth to derive sexual pleasure. Terminology is important here, because at least one team of psychiatrists writing on this subject distinguishes between autofellatio and self-irrumatio. In nonsolo sex, fellatio sees most of the action in the sucking party, while irrumatio has more of a thrusting element to it, wherein the other person's mouth serves as a passive penile receptacle. (Hence the colorful and rather aggressive-sounding slang for irrumatio— "face-fucking," "skullfucking," and so on.)

In any event, my paper-towel-roll act was simply a Plan B at that puerile age, a futile way to circumvent the obvious anatomical limitations to oral self-gratification. And by all accounts, I wasn't alone in hatching Plan A. Alfred Kinsey and his colleagues reported in *Sexual Behavior in the Human Male*, in fact, that "a considerable portion of the population does record attempts at self-fellation, at least in early adolescence." Sadly, given our species' pesky rib cage and hesitant spine, Kinsey estimated that only two or three of every one thousand males are able to achieve this feat. There's the story of the Italian decadent poet Gabriele D'Annunzio, who is said to have had a bone removed to facilitate the act, or that old *Saturday Night Live* skit in which Will Ferrell enrolls in a yoga class only to become flexible enough to fellate his own organ. But truth is often stranger than fiction. In 1975, the psychiatrist Frances Millican and her colleagues described the real case of a "very disturbed" patient who learned yoga precisely for this reason.

Now, you may think that being one of the ultrabendable percent of the population is all fun and games. (We've all heard those quips about never having to leave the house.) But think again. There's a long and unfortunate history of pathologizing this behavior; psychiatrists have described its practitioners as being sexually maladjusted, stuck in an infantile state of suckling dependency, or even motivated by repressed homosexual desires. Take the case described by the psychiatrists Jesse Cavenar, Jean Spaulding, and Nancy Butts, who wrote in 1977 of a lonely twenty-two-year-old serviceman who'd been fellating himself since the age of twelve. He was driven mad "by the fact that he could physically incorporate only the glans, and wanted to be able to incorporate more." Honestly, it must have been so—oh, what's the word I'm looking for . . . it's right on the tip of my tongue—*frustrating*, for this poor soldier. This is the ultimate cock tease, its being so close yet so far away.

Since the days of Freud, psychoanalysts have gone to town on the subject of autofellatio. In a 1971 article by the psychiatrist Frank Orland, we see the typical jargon-filled language used in dissecting the "symbolic" bases of autofellatio, which is conceptualized as a virtual "ring of narcissism":

> Autofellation represents a recreation of the early infantile state in which the intrapsychic representatives of external objects are separated from the self-object, with a coexisting parasitic symbiosis with the external object. Through the autofellatio phenomenon, the ego reestablishes the necessary mastery over the external object representative as a defence against object loss and to restore the parasitic fusion with the nipple-breast.

That, ladies and gentlemen, is unadulterated psychobabble—and I tell you this as a psychologist. Sometimes people are motivated to lick their own genitals because it just feels good. Of course, there are always going to be those, such as the dubious yoga master, who take it a bit too far and for whom autofellatio contributes to mental illness. The foregoing soldier, who couldn't take it far enough, got so frustrated by his semi-fulfilled fantasy that when he masturbated the old-fashioned way, he could achieve climax only by imagining himself fellating himself.

The very first published psychiatric case of autofellatio,

appearing in *The American Journal of Psychiatry* way back in 1938, was also one of the most outrageous and pathological. The patient was a thirty-three-year-old store clerk who, prior to being referred to the Yale psychiatrists Eugen Kahn and Ernest Lion, had just completed a sixty-day jail sentence for sexual assault. "Among his perverse practices," explain the authors, "were pedophilia, cunnilinguism, homosexual acts (fellatio, sodomy and mutual masturbation), exhibitionism, transvestism, fetishism, algolagnia, voyeurism and peeping." But never mind all those plain vanilla paraphilias. The man's psychiatrists were especially intrigued by his more unusual habit. He seems to have been a devious wee character, this patient of theirs. The authors describe him as being somewhat effeminate in posture, gait, and mannerisms; he stood only five feet two inches tall—"somewhat thin and with wide hips," they wrote, with "a female pattern of distribution of his pubic hair," and "his gag reflex is very sluggish."

The patient was the third oldest of eight children and grew up in a strict, religious family, which the physicians felt he rebelled against by egregiously breaching their high moral standards. In recounting to the psychiatrists the origins of his interest in autofellatio, the troubled clerk recalled being invited at the age of fourteen by a "cripple boy" to engage in oral sex with him. The patient, being shy, had refused this offer, but the thought of it simmered and, lacking the courage to approach anyone else, he took matters upon himself: "He kept trying night after night, managing to bend his back more and more until he finally succeeded in August, 1923." (Just in case you want to mark the anniversary on your calendar.) It turns out he liked it—so much, in fact, that even amid the long litany of perversions he enjoyed, self-irrumatio instantly became his favorite autoerotic act.

In an odd Pavlov's dog sort of way, the authors even describe how the man's sexual arousal had since then been accompanied by a "constricting feeling in the throat." That must be a terribly annoying feeling, I'd imagine, and apparently also one not easily resolved. "He has attempted to secure substitute gratification," say the authors, "by smoking, or by stimulating his pharynx with a banana, vaginal douche or a broom handle. These have yielded various degrees of satisfaction." And he did apparently get over his adolescent shyness and lack of confidence, too: he particularly enjoyed fellating himself in front of a shocked audience.

Since this initial case report by Kahn and Lion, a handful of others have trickled in over the years, with subsequent investigators attempting to find a set of common personality denominators in those who prefer autofellatio over other forms of sex. In a 1954 article in *Psychoanalytic Review*, for instance, William Guy and Michael Finn saw a theme beginning to emerge. "In all of the clinical descriptions," observe these authors, "one finds repeatedly such phrases as sensitive, shy, timid, effeminate, and passive." This is code for "queer," I believe, and in fact other writers have more expressly noted the often-suppressed homosexual desires in these autofellators.

In fact, judging by the scant literature, one of the big psychoanalytic questions yet to be resolved satisfactorily seems to be the extent to which engaging in autofellatio—or perhaps simply the desire to do so—signals a latent erotic attraction to the same sex. I suspect, however, that the overrepresentation of gay men in the antiquated case reports is simply a reflection of the cultural ethos of those times. The most recent psychiatric investigations on autofellatio date to the late 1970s (around the time that Freud's particular grip on psychiatry lost its tenuous hold), and the earlier ones to the 1930s, so as a rule the men described therein faced baseless moralistic proscriptions against homosexuality. This meant other men's penises were very hard to come by. So it's not terribly surprising that those too frightened or closeted to perform fellatio on another man would develop severe neuroses after indulging in their own penises.

A 1946 article from *The American Journal of Psychiatry* exemplifies this phenomenon. The case involves a thirty-six-year-old, highly intelligent, personable, but virginal staff sergeant (not to be confused with the military man we met earlier) with closeted homosexual desires. According to the official record, he'd first performed autofellatio at age thirteen, but he became so frightened by this "impulse" that he resisted ever doing so again—that is, until a month prior to arriving at the psychiatric ward of the hospital. After giving himself head in private, the sergeant became intensely paranoid that the other soldiers somehow knew of his autofellatio and that every little snigger, whisper, or averted glance concerned this transgression. He suffered a nervous breakdown on hearing the word "cocksucker" floating about so casually and playfully in the military barracks, convinced it was meant just for him.

It's a rather sad ending for him, too, because despite his responding well to the doctors' reassurance that he was being overly paranoid, the sergeant was discharged for being "no longer adaptable within the military service." The therapists assigned to the case, Major Morris Kessler and Captain George Poucher, reached a rather strange conclusion, one that I have a hunch you might disagree with. "Sexual self-sufficiency," they write, "either by masturbation or autofellatio, is tantamount to having an affinity for one's own sex." In other words, if you were a fan of manual masturbation in 1946, my heterosexual male friends, you'd have been branded a secret homosexual pervert who likes penises so much that he gives himself hand jobs. This would have made autofellatio a devil of a case under the Clinton-era "Don't Ask, Don't Tell" ban on gays in the military had it arisen then. And, seriously, good riddance to those ignorant days of yore. To each his own—quite literally in the case of autofellatio.

I know, I know, I didn't even get a chance to talk about autocunnilingus in females. Given the even more serious anatomical hurdles in lacking a protruding reproductive device, such behavior in women may not even be possible. I confess I don't know; and there's no mention of it in the scientific literature. The closest female comparison to autofellatio I stumbled upon is the case of women who suckle from their own *breasts*, for sexual or other purposes. One therapist writes of an especially self-sufficient female patient who had a habit of doing this. When he asked her why, she merely replied, "I'm hungry." But that's another story for another day.

Why Is the Penis Shaped Like That?
The Extended Cut

If you've ever had a good long look at the human phallus, whether yours or someone else's, you've probably scratched your head over its peculiar shape. Let's face it: it's not the most intuitively configured appendage in all of evolution. But according to the evolutionary psychologist Gordon Gallup, the human penis is actually an impressive "tool" in the truest sense of the word, one manufactured by nature over hundreds of thousands of years of human evolution. You may be surprised to discover just how highly specialized a tool it is. Furthermore, you'd be amazed at what its appearance can tell us about the nature of our sexuality.

The curious thing about the evolution of the human penis is that for something that differs so obviously in shape and size from that of our closest living relatives, only in the past few years have researchers begun studying its natural history in any detail. The reason for this neglect isn't clear. It's hard to imagine that hard-nosed scientists would be worried about the subject stirring up uncomfortable puritanical sentiments. The issue does have an intrinsic snicker factor, so I realize it takes a special type of psychological scientist to tell the little old lady sitting next to him on a flight to Denver that he studies how people use their penises when she asks what he does for a living. In any event, if you think there's only one way to use your penis, that it's merely an instrument of internal fertilization that doesn't require further thought, or that size doesn't matter, well, that just goes to show how much you can learn from Gallup's research findings.

Gallup's approach to studying the design of the human penis is a perfect example of *reverse engineering* as the term is used in the field of evolutionary psychology, and reverse engineering is an often unspoken concept that you'll find me using repeatedly throughout this book. This is a logico-deductive investigative technique for uncovering the adaptive purpose or function of existing (or extant) physical traits, psychological processes, or cognitive biases. That is to say, if you start with what you see today—in this case, the oddly shaped penis, with its bulbous glans (the "head," in common parlance), its long, rigid shaft, and the coronal ridge that forms a sort of umbrella-lip between these two parts—and work your way backward regarding how it came to look like that, the reverse engineer is able to posit a set of functionbased hypotheses derived from evolutionary theory. In the present case, we're talking about penises, but the logic of reverse engineering can be applied to just about anything organic, from the shape of our incisors to the opposability of our thumbs or the arch of our eyebrows.

For the evolutionary psychologist, the pressing questions are, essentially, *Why is it like that?* and *What is that for?* The answer isn't always that it's a biological adaptation—that it solved some evolutionary problem and therefore gave our ancestors a competitive edge in terms of their reproductive success. Sometimes a trait is just a "by-product" of other adaptations. Blood isn't red, for example, because red worked better than green or yellow or blue, but only because it contains the red hemoglobin protein, which happens to be an excellent transporter of oxygen and carbon dioxide. But in the case of the human penis, all signs point to a genuine adaptive reason that it has come to look the way it does.

If you were to examine the penis objectively—please don't do this in a public place or without the other person's permission—and compare the shape of this organ with the design of the same organ in other species, you'd notice the following uniquely humanf characteristics. First, despite variation in size between individuals, the human penis is especially large compared with that of other primates. When erect, it measures on average between five and six inches in length and about five inches in circumference. Even the most well-endowed chimpanzee, the species that is our closest living relative, doesn't come anywhere near this. Rather, even after correcting for overall mass and body size, chimp penises are about half the size of human penises in both length and circumference. I'm afraid that I'm a more reliable source on this than most. Having spent the first five years of my academic life studying great ape social cognition, I've seen more simian penises than I care to mention. I once spent a summer with a 450-pound silverback gorilla that was hung like a wasp (great guy, though) and babysat a lascivious young orangutan that liked to insert his penis in just about anything with a hole, which unfortunately one day included my ear.

In addition, only the human species has such a distinctive mushroom-capped glans, which is connected to the shaft by a thin tissue of frenulum (the delicate tab of skin just beneath the urethra). Chimpanzees, gorillas, and orangutans have a much less extravagant phallic design— more or less all shaft. It turns out that one of the most significant features of the human penis isn't so much the glans per se as the coronal ridge it forms underneath. The diameter of the glans where it meets the shaft is wider than the shaft itself. This results in the coronal ridge that runs around the circumference of the shaft—something Gallup, by using the logic of reverse engineering, believed might be an important evolutionary clue to the origins of the strange sight of the human penis.

Now, the irony doesn't escape me. But even though this particular evolutionary psychologist (yours truly) is gay, for the purposes of research we must consider the evolution of the human penis in relation to the human vagina. Magnetic imaging studies of heterosexual couples having sex reveal that during coitus, the typical penis completely expands and occupies the vaginal tract and with full penetration can even reach the woman's cervix and lift her uterus. This, combined with the fact that human ejaculate is expelled with great force and over considerable distance (up to two feet if not contained), suggests that men are designed to release sperm into the uppermost portion of the vagina possible. In an article published in the journal *Evolutionary Psychology*, Gallup and Rebecca Burch argue that "a longer penis would not only have been an advantage for leaving semen in a less accessible part of the vagina, but by filling and expanding the vagina it also would aid and abet the displacement of semen left by other males as a means of maximizing the likelihood of paternity."

This "semen displacement theory" is the most intriguing part of Gallup's story. We may prefer to regard our species as being blissfully monogamous, but at least some degree of fooling around has been our modus operandi for at least as long as we've been on two legs. Since sperm cells can survive in a woman's cervical mucus for up to several days, if she has more than one male sexual partner over this period of time, say within forty-eight hours, then the sperm of these two men are competing for reproductive access to her ovum. According to Gallup and Burch, "Examples include group sex, gang rape, promiscuity, prostitution, and resident male insistence on sex in response to suspected infidelity." And although semen displacement is the competing male's goal, even nicely evolved penises aren't perfect. In fact, the authors cite the well-documented cases of human heteroparity, where "fraternal twins" are in fact sired by two different fathers who had sex with the mother within close succession of each other, as evidence of our species' natural sexual inclinations.

So how did nature equip men to solve the adaptive problem of other men impregnating their sexual partners? The answer, according to Gallup, is that their penises were sculpted in such a way that the organ would effectively displace the semen of competitors from their partner's vagina, a well-synchronized effect facilitated by the "upsuck" of thrusting during intercourse. Specifically, the coronal ridge offers a special removal service by expunging foreign sperm. According to this analysis, the effect of thrusting would be to draw other men's sperm away from the cervix and back around the glans, thus scooping out the semen deposited by a sexual rival.

You might think this is all fine and dandy, but one can't possibly prove such a thing. You'd be underestimating Gallup, however, who just so happens to be a very talented experimental researcher (among other things, he's also well known for developing the famous mirror self-recognition test for use with chimpanzees back in the early 1970s). In a series of studies published in *Evolution and Human Behavior*, Gallup and a team of his students put

the semen displacement hypothesis to the test using
artificial human genitalia of different shapes and sizes.
They even concocted several batches of realistic seminal
fluid.

Findings from the study may not have "proved" the
semen displacement hypothesis, but they certainly confirmed its principal points. Here's how the basic study
design worked. (And perhaps I ought to preempt the usual
refrain by pointing out that yes, Gallup and his coauthors
did receive full ethical approval from their university to
conduct this study.) The researchers selected several sets of
prosthetic genitals from erotic novelty stores, including a
realistic latex vagina, sold as a masturbation pal for lonely
straight men and tied off at one end to prevent leakage,
and three artificial phalluses. The first latex phallus was
6.1 inches long and 1.3 inches in diameter with a coronal
ridge extending approximately 0.20 inch from the shaft.
The second phallus was the same length, but its coronal
ridge extended only 0.12 inch from the shaft. Finally, the
third phallus matched the other two in length but lacked a
coronal ridge entirely. In other words, whereas the first
two phalluses closely resembled an actual human penis,
varying only in the coronal ridge properties, the third (the
control phallus) was the bland and headless horseman of
the bunch.

Next, the researchers borrowed a recipe for simulated
semen from another like-minded evolutionary psychologist, Todd Shackelford, and created several batches
of seminal fluid. The recipe "consisted of 0.08 cups of
sifted, white, unbleached flour mixed with 1.06 cups
of water. This mixture was brought to a boil, simmered for
15 minutes while being stirred, and allowed to cool." In a
controlled series of "displacement trials," the vagina was
loaded with this fake semen, and the phalluses were

inserted at varying depths (to simulate thrusting) and removed, whereupon the latex orifice was examined to determine how much semen had been displaced from it. As predicted, the two phalluses with the coronal ridges displaced significantly more semen from the vagina (each removed 91 percent) than the "headless" control (35.3 percent). Additionally, the farther the phalluses were inserted—that is to say, the deeper the thrust—the more semen was displaced. When the phallus with the more impressive coronal ridge was inserted three-fourths of the way into the vagina, it removed only a third of the semen, whereas it removed nearly all of the semen when inserted completely. Shallow thrusting, simulated by the researchers inserting the artificial phallus halfway or less into the artificial vagina, failed to displace any semen at all. So if you want advice that'll give you a leg up in the evolutionary arms race, don't go west, young man—go deep.

For the second part of the study, Gallup administered a series of survey questions to college-age students about their sexual history. Drawing from previous studies that showed how sexual jealousy inspires predictable (and biologically adaptive) "mateguarding" responses in human males, these questions were meant to determine whether certain "penile behavior" (my term, not theirs) could be expected based on the men's suspicion of infidelity in their partners. In the first of these anonymous questionnaires, heterosexual men and women reported that in the wake of allegations of female cheating, men thrust deeper and faster. Results from a second questionnaire revealed that upon first being sexually reunited after time apart, couples engaged in more vigorous sex—namely, compared with baseline sexual activity where couples see each other more regularly, vaginal intercourse following periods of separation involved deeper and quicker thrusting. Hopefully, you're thinking as an evolutionary psychologist at this point and can infer what these survey data mean: by using their penises proficiently as a semen displacement device, men are subconsciously (in some cases consciously) combating the possibility that their partners have had sex with another man in their absence.

Doubtful about this interpretation? The really beautiful thing about evolutionary psychology—or the most frustrating, if you're one of its many critics—is that you don't have to believe it's true for it to work precisely this way. Natural selection doesn't much mind if you favor an alternative explanation for why you get so randy upon being reunited with your partner. Your penis will go about its business of displacing sperm regardless.

There are many other related hypotheses that are based on the central logic of the semen displacement theory. In their 2004 *Evolutionary Psychology* piece, for example, Gallup and Burch expound on a number of fascinating spin-off ideas in a followup article to their earlier work on the natural history of the penis. For example, one obvious criticism of the theory is that men would essentially disadvantage their own reproductive success by removing their own sperm cells from their sexual partner. However, in your own sex life, you've probably noticed the "refractory period" immediately following ejaculation, during which males almost instantly lose their tumescence (the erection deflates to half its full size within one minute of ejaculating), their penises become rather hypersensitive, and further thrusting even turns somewhat unpleasant. In fact, for anywhere between thirty minutes and twenty-four hours, most men are rendered temporarily impotent following ejaculation. According to Gallup and Burch, these post-ejaculatory features, in addition to the common "sedation" effect of orgasm, may be adaptations to the problem of "self-semen displacement," meaning that the odds of removing your own sperm are reduced considerably when your penis is sore or flaccid or while you're soundly asleep.

Gallup and Burch also leave us with a very intriguing hypothetical question in their article. "Is it possible (short of artificial insemination)," they ask, "for a woman to become pregnant by a man she never had sex with? We think the answer is 'yes.' " It's a tricky one to wrap your head around, but basically Gallup and Burch say that semen displacement theory predicts that something like the following example would be possible (note that I've modified this from the original article for your reading pleasure; also, observe how the scenario is especially relevant to uncircumcised men): If "Josh" were to have sex with "Kate," who recently had sex with "Mike," in the process of Josh's thrusting his penis back and forth in her vagina, some of Mike's semen would be forced under Josh's frenulum, would collect behind his coronal ridge, and would be displaced from the area proximate to the cervix. After Josh ejaculates and substitutes his semen for that of the other male, as he withdraws from the vagina, some of Mike's semen will still be present on the shaft of his penis and behind his coronal ridge. As his erection subsides, the glans will withdraw under the foreskin, raising the possibility that some of Mike's semen could be captured underneath the foreskin and behind the coronal ridge in the process. Were Josh to then have sex with "Amy" several hours later, it is possible that some of the displaced semen from Mike would still be present under his foreskin and thus may be unwittingly transmitted to Amy, who in turn could then be impregnated by Mike's sperm . . .

It's not exactly an immaculate conception. But just imagine the gasps from your average Maury Povich show audience.

People have some pretty strong feelings about penises. Initial reactions to the essay you've just read ranged from the incredulous ("Are you seriously suggesting that chimpanzees aren't promiscuous?"), to the imaginative ("Penises! They're so cute, you just want to pinch their cheeks and give them cookies"), to the rather irritable ("Stupid, biased thinking again from an 'evolutionary psychologist' "). So I decided to speak directly with Gordon Gallup, whose controversial semen displacement theory, after all, was the one that had struck up such a brouhaha regarding the adaptive functioning of this enigmatic organ. Perhaps, I appealed to him, he might offer us a few more clarifying details regarding the theory. I took many of the "core" questions to heart and asked Gordon to respond to several of them:

READERS: The latex genitalia study wasn't terribly convincing because the models were circumcised, and in real life the foreskin would interfere with the semen-displacing functions of the coronal ridge. So, does the foreskin pose a problem for the semen displacement theory?

GALLUP: The length of the foreskin is one of the most variable features of the human penis. When most uncircumcised males achieve an erection it pulls the foreskin back over the glans and back down the shaft of the penis, enabling the coronal ridge to do its business and scoop rival males' semen away from the woman's cervix. Because circumcision reduces the diameter of the shaft immediately behind the glans and accentuates the coronal ridge, we've speculated that the practice of circumcision may have unwittingly modified the penis in ways that enable it to function as a more effective semen displacement device. Armchair speculation? No. The idea could be tested by comparing the incidence of non-paternity between circumcised and intact males. My prediction would be that circumcised males ought to experience a lower incidence of being cuckolded.

READERS: So why did human penises evolve to have foreskin at all then?

GALLUP: Evolution does not occur by design. The best way to think about most adaptations is in terms of cost/benefit ratios. I suspect that the foreskin provided protection of the glans and what you see is the result of a statistical compromise of sorts.

READERS: If the penis really evolved to displace semen, then why wouldn't other promiscuous primate species, namely chimpanzees, have evolved similarly designed penises with the coronal ridge?

GALLUP: Again, evolution doesn't occur by design. It occurs by selection, and the raw material for such selection consists of nothing more than random genetic accidents (mutations). Embedded in the evolutionary history of human genital design were some penis shape mutations, not present in other species, that led to a device that could be used to compete with other males for paternity. Other promiscuous primates such as chimpanzees have solved the problem through sperm competition. Male chimpanzees have testicles that are three times the size of humans and differences in sperm count are on the same order of magnitude. Chimpanzees compete among one another for paternity by leaving the largest and most potent volume of semen in the female reproductive tract. When it comes to selection based on genetic accidents, there are a number of ways to skin the adaptive cat.

Bering here. And speaking of cats, and penises, it's perhaps useful to reflect in closing on cat penises. Like human males, male cats possess remarkably specialized penises. They come equipped with a band of about 150 sharp, backward-pointing spines that, literally, rake the internal walls of the female cat's vagina (hence the deafening yowl that often accompanies feline sex). This both triggers ovulation and displaces the sperm of prior males that may have recently mounted her. We should give thanks—and I say this as a gay man, and one not without some stakes in this whole painful affair—that evolution took a somewhat gentler course in our species.

Not So Fast . . . What's So "Premature" About Premature Ejaculation?

It occurred to me recently, under conditions that I leave to your ample and likely sordid imagination (how *dare* you), that the very concept of premature ejaculation in human males is a strange one, at least from an evolutionary theoretical perspective. After all, the function of ejaculation isn't really a mysterious biological occurrence: it's an evolved mechanism designed by nature to launch semen, and therefore sperm cells, as far into the dark, labyrinthine abyss of the female reproductive tract as possible. And once one of these skyrocketed male gametes, in a vigorous race against millions of other single-tasked cells, finds and penetrates a fertile ovum, and—miracle of miracles—successful conception occurs, well, then, natural selection can congratulate itself on a job well done.

So given these basic biological facts, and assuming that ejaculation is not so premature that it occurs prior to intromission and sperm cells find themselves awkwardly outside of a woman's reproductive tract flopping about like fish out of water, what, exactly, is so "premature" about premature ejaculation? In fact, all else being equal, in the ancestral past, wouldn't there likely have been some reproductive advantages to ejaculating as quickly as possible during intravaginal intercourse—such as inseminating as many females as possible in as short a time frame as possible? Or allowing our ancestors to focus on other adaptive behaviors aside from sex? Or perhaps, under surreptitious mating conditions, doing the deed quickly and expeditiously without causing a big scene?

Like so many things before, it turns out that this insight of mine was actually several decades behind the curve, because in 1984, when, at nine years of age, I was still anything but a premature ejaculator, a sociologist named Lawrence Hong published a highly speculative but very original paper along these same lines that's worth engaging with here, fittingly titled "Survival of the Fastest: On the Origin of Premature Ejaculation." In this article, Hong—whose most recent work, as far as I can tell, has been on the global phenomenon of cabaret transgenderism—posited that during the long course of human evolutionary history, "an expeditious partner who mounted quickly, ejaculated immediately, and dismounted forthwith might [have been] the best for the female."

The empirical centerpiece of Hong's arriving at this conclusion is the fact that on average, human males achieve orgasm by ejaculating just two minutes after vaginal penetration, whereas it takes the owners of these vaginas, on average, at least twice that long to do the same once a penis is inside of them—if they achieve orgasm at all, that is. This obvious gender mismatch between orgasm latencies can be understood, Hong reasons, only once we acknowledge that sex evolved, at least initially, for purely reproductive purposes. Don't forget, he reminds us, that recreational heterosexual sex is enabled only by relatively recent technological innovations, such as contraceptive devices.

Hong compares the mating habits of human beings with

those of other rapid—and not-so-rapid—ejaculators in the primate family, noting that the faster a primate species is in the coital realm, the less aggressive it is when it comes to mating-related behaviors. He calls this the "slow speed–high aggressiveness hypothesis." For example, male rhesus macaque monkeys often engage in marathon mounting sessions, where sex with a female can be drawn out for over an hour at a time (including many breaks and therefore noncontinuous thrusting). That may sound great, but libidinous anthropomorphizers beware: macaque sex is a chaotic and violent affair, largely because the duration of the act often draws hostile attention from other competitive males. By contrast, primate species whose males evolved to ejaculate rapidly would have largely avoided such internecine violence, or at least minimized it to a considerable degree.

Key to Hong's analysis is the idea that intravaginal ejaculation latencies in males are heritable; there was initially greater within-population-level variation in our male ancestors, he surmises, but over time "the ancestry of *Homo sapiens* became overpopulated with rapid ejaculators." According to Hong, this is because young reproductive-aged males who ejaculated faster (that is, had more sensitive penises) avoided injury, lived longer, and therefore had a greater chance of attaining high status and acquiring the most desirable females.

Hong's reasoning on these heritability grounds has in fact received recent support. You may have missed this in your monthly periodical readings, but in a 2009 article from *The International Journal of Impotence Research* a team of Finnish psychologists led by Patrick Jern reported evidence from a large-scale twin study showing that premature ejaculation is determined significantly by genetic factors. Thousands of male twin pairs—fraternal and identical—completed a survey about how long it took them to reach orgasm; and the timing of identical twins was more closely matched than fraternal twins. So just as Hong surmised many years ago, this is indeed a heritable trait; if you doubt it, go on, have that awkward conversation with your fathers, boys. In fact, since Jern and his colleagues found that delayed ejaculation—the other extreme end of the ejaculation latency continuum—revealed no such genetic contributions, these authors generally agree with Hong, postulating that "premature" ejaculation may be a product of natural selection whereas delayed ejaculation "would be completely maladaptive." Delayed ejaculators are considerably rarer, with a prevalence rate as low as 0.15 percent in the male population compared with as high as 30 percent with premature ejaculators, and their condition is usually owed to lifelong medical conditions or the recent use of anti-adrenergics, selective serotonin reuptake inhibitors, neuroepileptics, or other modern-day drugs that are often associated with anorgasmia as a miserably unfortunate side effect.

Adding further credence to the evolutionary model is a separate set of self-report data published in *The Journal of Sexual Medicine*, in which Jern and colleagues demonstrated that ejaculation latencies were significantly shorter when men achieve orgasm through vaginal penetration than when doing so in the course of other activities, such as anal, oral, or manual sex. In fact, in light of these differential ejaculation latencies, they argue that the very construct of such male orgasmic "timing" is best carved up by discrete sexual behaviors rather than treated as a more general clinical phenomenon. And they offer several helpful acronyms for these ejaculation latency subtypes, too, such as "OELT" for "oral ejaculation latency time" and, conveniently, "MELT" for "masturbation ejaculation latency time."

I have the niggling, faraway sense that we've left something out of the evolutionary equation regarding the variation in male ejaculation latencies. What, oh what, can that possibly, conceivably be? Oh, come now, I know it's *women's sexual satisfaction*. Actually, Hong didn't leave female orgasms out of his rather viscous analysis altogether; he just didn't see them as being central to selective pressures. Presumably, like other theorists of that time writing about the biological reasons for female orgasms (such as Stephen Jay Gould, who thought that female orgasms were much like male nipples, a happy leftover of the human embryological *Bauplan*), he saw women's sexual pleasure as being a nice, but neither here nor there, feature of human sex that nature had thrown into the mix.

Hong acknowledges—with great humility and humor, in fact—that his ideas on the evolutionary origins of premature ejaculation in human males are mostly guesswork. And his ideas were critiqued by the psychologist Ray Bixler in his review of Hong's theory. Among many faults that Bixler finds in Hong's "survival of the fastest" theory, the basic logic just doesn't mesh with the obvious female pursuit of sexual intercourse. In chimpanzees, for instance—a species for which male ejaculation latencies are measured in seconds, not minutes—it is often females who initiate mating behaviors. And then there's the "ouch" factor of having a nonaroused female partner whose dry genitals aren't terribly inviting. If Hong's model were correct, says Bixler, "there would be little or no proximal cause, other than coercion, for female cooperation—and it should be very clear that she would have to cooperate if voluntary mating were to be speedy! If she were not lubricated he would have 'to rasp it in,' a painful experience for the woman, and . . . 'no pleasure' for him either."

Disappointingly, this is more or less where the evolutionary thinking stops on this subject. Apparently, no other theorist—at least no experimentally inclined evolutionary theorist—has picked up Hong's lead in trying to tease apart competing adaptationist arguments regarding male ejaculation latencies. Pieces of the puzzle are floating about out there, I suspect, such as the Finnish research showing that vaginal sex leads to faster ejaculations compared with other sexual behaviors. But Hong's article was before its time—premature itself, in light of today's more informed evolutionary biology, which is now poised to construct a more nuanced empirical model about this evolutionary legacy that is behind so many of us being fast finishers.

Another big piece of the puzzle may be linked to our species' uniquely evolved social cognitive abilities. Possibly only tens of thousands of years ago, just a splinter of a splinter's time in the long course of our primate history, ancestral humans may have become the only species capable of experiencing empathy with our sexual partners during intercourse. Men could then think about satisfying their partners during sex rather than just themselves, thus deliberately prolonging the act of coitus to delay their own orgasm for her sake. Prior to this, our more distant ancestors may have been more like chimpanzees, seeing others' bodies as mindless meat.

Given the unpleasant stigma attached to premature ejaculation, an evolutionary approach to the "problem" could greatly inform clinical treatments, a (not surprisingly) high-grossing therapeutic area in which there is no shortage of work being done. But in any event, Hong's seminal ideas should give us all pause in labeling any particular intravaginal ejaculation "premature"; Mother Nature, arguably the only lover that really matters, after all, may very well have had a thing for our one-minute ancestors.

An Ode to the Many Evolved Virtues of Human Semen

I have come upon a secret treasure, a heretofore-unknown bounty of facts only recently unearthed by a team of evolutionary psychologists. A vital forewarning, though: although the data and information I am about to share ooze with the promise of dramatically improving virtually every aspect of your well-being, it can also be abused with tragic—even fatal—consequences. This is so much the case, in fact, that I debated the merits of popularizing this material and do so here only with great circumspection and caution. So please be wise in digesting this semen-related knowledge, and be wiser still in applying it to your own sex lives.

As with the origins of so many great scientific discoveries, this story begins with a serendipitous chain of events. "Our interest in the psychological properties of semen arose as a by-product of an initial interest in menstrual synchrony," explain the codiscoverers Gordon Gallup and Rebecca Burch, writing about human semen. In particular, Gallup and Burch had stumbled onto a set of curious data from the mid-1990s showing that unlike heterosexually active women residing together, partnered lesbians sharing a residence failed to exhibit the well-known "McClintock effect," in which menstrual cycles in cohabiting women (as well as those of females from many other species) are synchronized. Since subtle olfactory cues (called pheromones) are known to mediate menstrual synchrony, write the authors, "this struck us as peculiar . . . because lesbians would be expected to be in closer, more intimate contact with one another on a daily basis than other females who live together. What is it about heterosexual females that promotes menstrual synchrony, or conversely what is it about lesbians that prevents menstrual synchrony? It occurred to us that one feature that distinguishes heterosexual women from lesbians is the presence or absence of semen in the female reproductive tract. Lesbians have semen-free sex."

Perhaps you already see where this is leading. Gallup and Burch reasoned that certain chemicals in human semen, through vaginal absorption, affect female biology in such a way that women who have condom-less sex literally start to smell different from those women—lesbians or otherwise—who do not. At least, bodies of the former emit pheromones that "entrain" menstrual cycles among cohabiting women. (Their hunch was indeed borne out as they reviewed the existing literature on menstrual synchrony.) But this happenstance discovery of asynchronous lesbians was just the tip of the semen iceberg for Gallup and Burch, who quickly discovered that although much was known among biologists about basic semen chemistry, virtually nothing was known about precisely how these chemicals might influence female biology, behavior, and psychology.

And that is a rather odd omission in the biological literature indeed, since there could hardly be anything more obvious in Darwinian terms than the fact that semen is, almost by definition, naturally designed to get into the chemically absorptive vagina. Bear in mind that although they are often conflated in everyday parlance, along with many other less scientific terms, semen is not the same thing as sperm. In fact, you may be surprised to learn that only about 1 to 5 percent of the average human ejaculate consists of sperm cells. The rest of the ejaculate, once drained of these famously flagellating gametes, is referred to as "seminal plasma." So when one discusses the chemical composition of semen, it is the plasma itself, not the spermatozoa, that is at issue.

Now, medical professionals have known for a very long time that the vagina is an ideal route for drug delivery. This is because the vagina is surrounded by an impressive vascular network. Arteries, blood vessels, and lymphatic vessels abound, and—unlike some other routes of drug administration—chemicals that are absorbed through the vaginal walls have an almost direct line to the body's peripheral circulation system. So it makes infinite sense, argue Gallup and Burch, that like any artificially derived chemical substance inserted into the vagina via pessary, semen might also have certain chemical properties that tweak female biology.

It turns out that this insight, so obvious as to be all but invisible, has been a theoretical gold mine for this hawkeyed pair of adaptation-minded thinkers. But before we jump into their rich vat of seminal theory, let's have a quick look at some of the key ingredients of human semen. In fact, semen has a very complicated chemical profile, containing more than fifty different compounds (including hormones, neurotransmitters, endorphins, and immuno-suppressants), each with a special function and occurring in different concentrations within the seminal plasma. Perhaps the most striking of these compounds is the bundle of mood-enhancing chemicals in semen. There is good in this goo. Such anxiolytic chemicals include, but are by no means limited to, cortisol (known to increase affection), estrone (which elevates mood), prolactin (a natural antidepressant), oxytocin (also elevates mood), thyrotropin-releasing hormone (another antidepressant), melatonin (a sleep-inducing agent), and even serotonin (perhaps the best-known antidepressant neurotransmitter).

Given these ingredients—and this is just a small sample of the mind-altering "drugs" found in human semen—Gallup and Burch, along with the psychologist Steven Platek, hypothesized rather boldly that women having unprotected sex should be less depressed than suitable control participants. To investigate whether semen has antidepressant effects, the authors rounded up 293 college females from the SUNY Albany campus who agreed to fill out an anonymous questionnaire about various aspects of their sex lives. Recent sexual activity without condoms was used as an indirect measure of seminal plasma circulating in the woman's body. Each participant also completed the Beck Depression Inventory, a commonly used clinical measure of depressive symptoms.

The most significant findings from this study, published with criminally modest fanfare in the *Archives of Sexual Behavior*, were these: even after adjusting for frequency of sexual intercourse, women who engaged in sex and "never" used condoms showed significantly fewer depressive symptoms than did those who "usually" or "always" used condoms. Importantly, these chronically condom-less, sexually active women also evidenced fewer depressive symptoms than did those who abstained from sex altogether. By contrast, sexually active heterosexual women, even really promiscuous women, who used condoms were just as depressed as those practicing total abstinence. In other words, it's not just that women who are having sex are simply happier, but happiness appears to be a function of the ambient seminal fluid pulsing through one's veins.

Relax, settle down, take a deep breath—I know what you're thinking. This is a correlation study, and there are scores of other possible causes and explanations, both those that the authors anticipated and controlled for in this study design (by all means read the original work for more details, but note that these between-group differences in depression panned out even after controlling for the use of oral contraceptives, days since last sex, frequency of sex, and duration of the relationship with the male partner) and probably some that you can come up with on your own. Even the authors urge some degree of skepticism: "It is important to acknowledge that these data are preliminary and correlational in nature, and as such are only suggestive. More definitive evidence for antidepressant effects of semen would require more direct manipulation of the presence of semen in the reproductive tract and, ideally, the measurement of seminal components in the recipient's blood."

Now, I'm hedging here, but one thing I do want to mention, with a helpful nod from the authors of this study, is that the antidepressant effects of seminal plasma may not be limited to vaginal absorption of its mood-brightening chemical properties. "It would be interesting to investigate," write Gallup and his coauthors, "the possible antidepressant effects of oral ingestion of semen, or semen applied through anal intercourse (or both) among both heterosexual couples as well as homosexual males." So in my plumbing of the empirical literature for studies on unprotected anal sex among gay males, otherwise known as "barebacking," I came across a load of research on this very topic. Most of this work is couched, understandably so, in the HIV-prevention literature. One particularly telling study, though, comes from a 2005 report from the journal *Nursing Inquiry* in which the Canadian investigators Dave Holmes and Dan Warner interviewed barebacking gay males—not while they were engaged in the act, but through later introspection—about their motivations for preferring unprotected anal sex over condoms in light of the obvious dangers of infection. The most intriguing result to emerge from this study, in the context of Gallup and Burch's overall theoretical perspective regarding the psychobiology of semen, was that so many of the barebacking interview subjects viewed the exchange of semen through unprotected anal sex as providing them with a palpable sense of "connectedness" with their same-sex partners, one that happened only with the internally unimpeded ejaculation.

Unfortunately, rather than investigate the possible psychobiological effects of semen exchange in this dynamic, Holmes and Warner leer through a fairly typical postmodernist lens to explore the symbolic nature of semen exchange in barebackers. Now, I ask you, which is the more informative paradigm for understanding why gay men would practice unsafe sex through unprotected anal intercourse: an evolutionary biological account taking into consideration the chemical composition of seminal plasma and its possible effects on attachment among gay men, or a symbolic, postmodernist perspective like the following one advanced by Holmes and Warner (in all fairness, this is just a snippet, but a good taste of their approach)?

The body becomes the locus of never-ending fights, a carnal battlefield. The escape route (lines of flight) is intrinsic to the deterritorialization of the Body-without-Organs through which one becomes someone else.

However, the lines of flight could have paradoxical effects. Indeed, they can be avenues of creative potential or, conversely, paths of great danger. Yet, it is "always in a line of flight that we create" . . . "that we must continue to experiment with such lines." Lines of flight (nuclei of resistance of resingularization and heterogenesis) permit freedom to surge through a process of creative transformation and metamorphosis.

Trust me, even in context, that reads as if the authors were cobbling together a Braille sentence using the random distribution of acne on someone's back. Sorry to sound a bit testes—er, testy—but while such soupy postmodernist rhetoric may still have its place in certain scholarly circles, when one is dealing with something as clinically important as unprotected sex among vulnerable populations, a scientific understanding of these people's motivations is essential before any intervention of their high-risk behaviors can even begin to occur.

You may also be beginning to realize the dangers that I alluded to at the start of this essay. For both men and women, heterosexual and homosexual, knowing that the penis is capable of dispensing a sort of natural Prozac—whether obtained vaginally, anally, or orally—without also considering the viral arms race involving sexually transmitted infections can lead to very tragic decisions indeed and many undocumented high-risk, private bedroom semen "experiments." But here's just one reason to put the brakes on such plans: the HIV virus, which evolved long after these adaptive antidepressant factors, has apparently come to pirate human semen, such that certain protein factors in seminal plasma, particularly a protein called prostatic acid phosphatase, make HIV up to 100,000-fold more potent than it is outside of the plasma.

In any event, Gallup and Burch's model also reminded me of those oft-cited Papua New Guinea tribes, such as the Sambia, and their semen-ingestion rituals involving young boys. On the surface, there's a puzzling scenario here: such cultures have long histories of being embroiled in violent warfare, and thus they tend to place extraordinarily high value on expressed masculinity. Yet ritualized homoerotic practices involving young boys fellating older males in order to ingest their semen are common. In an issue of the *Archives of Sexual Behavior*, Gilbert Herdt, a cultural anthropologist who studied the Sambia, along with his colleague Martha McClintock (the same McClintock after whom the menstrual cycle synchrony effect discussed earlier was named), describes how "by the age of 11–12 [Sambia] boys have become aggressive fellators who actively pursue semen to masculinize their bodies."

In the past, cultural anthropologists such as Herdt have conceptualized this semen ritual mostly in symbolic terms. Yet since testosterone from the seminal plasma could penetrate the oral mucosa, along with a surfeit of other hormones and chemicals having possible spin-off effects on male behavior, I do not find it inconceivable that there may be genuine psychobiological consequences of semen intake occurring in these young swallowing males that are not wholly out of line with the Sambia's own folk beliefs. It might not be a theory you want to run by your local pastor or bring up at your next PTA meeting, but you get the idea.

But let's get back to more everyday semen ingestion. (Maybe not every day, but you know what I mean.) In addition to their semen-as-antidepressant model, Gallup and Burch have worked out many other intricate, persuasive arguments about how the various chemicals in human semen served—and continue to serve—biologically adaptive functions for both sexes. For example, among the more curious ingredients in human semen are follicle-stimulating hormone (FSH) and luteinizing hormone (LH). This is curious, Gallup and Burch point out, because these are distinctively female hormones. What are female hormones doing in human semen? The authors speculate that the presence of FSH and LH in human semen is related to concealed ovulation in human females.

Unlike females of other primate species, women do not have breeding patterns governed by season or standardized cycles, and there are no obvious signals—such as a fire-engine-red swollen rear end—giving away their time of the month. So for a naive human male, impregnating a woman as a consequence of sexual intercourse is much more a roll of the dice than it is for males of other species in their mating behaviors. Just as with any other species, though, getting the timing right so that release of semen coincides with the release of eggs is key. As a counter-defense against women's concealed ovulation, male evolution had a trick up its sleeve, which was the ability to manipulate the timing of a woman's ovulation to suit a man's own insemination schedule; that is to say, semen chemistry gives premature eggs a nice little nudging. Hence the conspicuous presence of FSH (which causes an egg in the ovary to ripen and mature) and LH (which triggers ovulation and release of that egg).

In support of this theoretical claim about semen chemistry and concealed ovulation in human females, consider that chimpanzee semen lacks the FSH hormone altogether and the presence of LH is rather negligible, which makes sense, of course, since chimpanzees are cyclical breeders and ovulating females display their own personal red-light-district signs by way of swollen, multicolored anogenital regions. "Thus it would appear," reason Gallup and Burch, "that the chemistry of human semen has been selected to mimic the hormonal conditions that control ovulation, and as such may account for instances of induced ovulation (ovulation triggered by copulation at points in the menstrual cycle when ovulation would otherwise be unlikely)."

Believe it or not, we've only scratched the surface of the evolved-semen literature. Here's a snapshot of other recent findings from Gallup's lab, most, remember, needing further investigation before we jump to any strong conclusions: semen-exposed women perform better on concentration and cognitive tasks; women's bodies can detect "foreign" semen that differs from their long-term or recurrent sexual partner's signature semen (an evolved system that, Gallup believes, often leads to unsuccessful pregnancies—via greater risk of preeclampsia—because it signals a disinvested male partner who is not as likely to provide for the offspring); women who had unprotected sex with their partners—and therefore were getting regularly inseminated by them—experience more significant depression on breaking up with these men than those who were not as regularly exposed to an ex's semen (and they also go on the rebound faster in seeking new sexual partners, which presumably would help fix their semen-deprived depression). And the list goes on.

Before I bid adieu, please accept, ladies, in all sincerity, my humblest apologies for what is likely to be a flood of inappropriate remarks from men saying, "I'm not a medical doctor, but my testicles are licensed pharmaceutical suppliers." I'm just the merry messenger.

PART II

Bountiful Bodies

The Hair Down There: What Human Pubic Hair Has in Common with Gorilla Fur

Like many people, I ask myself continuously about some of life's biggest mysteries. Why are we here? What is the meaning of life? Why do we have those strangely sparse, wiry little hairs growing around our genitals—hair that is singularly different from all the other hair on our bodies? Fortunately, scientists have managed to put my mind at rest on at least one of these daunting existential questions. In recent years, it seems, researchers have made some tremendous advances in the study of pubic hair.

So, let's start with what we already know about pubic hair. It's a signature of sexual maturity, sprouting up around our groins sometime in early adolescence. If it appears on a person's body any time earlier than this in development (say, prior to the age of nine years old), something is clearly the matter. Some things just don't go together in this world—babies and pubic hair are definitely two of them.

Precocious puberty is no laughing matter, of course, because children who begin to develop secondary sexual characteristics unusually early in their development may in fact have some significant underlying health problem, such as a lesion on the central nervous system that prematurely activates the hypothalamus. But for one young couple in Alabama, the term "precocious puberty" hardly does justice to what they were observing with their infant son a few years ago. Just imagine changing your six-month-old's diapers and noticing what appears to be a tuft of light-colored pubic hair on his groin. Over the next ten months, the pubic hair would become progressively darker and adultlike, which—when accompanied by an oddly large penis for a sixteen-month-old and, ahem, frequent erections—was *finally* enough to prompt this pair to seek medical advice.

This was the background of the case as it was presented to a group of physicians that eventually reported it in *Clinical Pediatrics*. Upon examining the child, Samar Bhowmick and his colleagues noted with some astonishment that "the pubic hair was [that of an adolescent], mostly around the base of the phallus and was dark and curled." Further inspection revealed a healthy, bouncing baby boy—completely age appropriate in all other respects—but the laboratory results indicated an abnormally high level of testosterone. Eventually, the doctors cracked the case. It turned out that the boy's father had been spreading testosterone gel twice daily over his shoulders, back, and chest, having been prescribed this treatment by his doctor to treat a low libido brought on by depression. Because the little boy slept in the same bed as his parents, with his father cuddling and hugging his child just after applying the gel, the bare-skin contact was causing his son to become a man much earlier than nature intended. (A follow-up visit later revealed, fortunately, that the pubic hair had all but disappeared once the father was made aware of this effect of his gel use, and the doctors were hopeful that the child would have no long-term complications from the testosterone exposure.)

This peculiar case of the pubic-haired infant is so striking, obviously, because this type of distinctive groin pelage *tends* to coincide with sexual maturation, not the developmental stage in which we're just learning how to walk. The case also highlights the oddity that is human pubic hair more generally. After all, we appear to be the only species of primate (perhaps the only species, period) that bears this type of strange hair around our genitals. Robin Weiss, a researcher from University College London's Division of Infection and Immunity, found himself standing in the shower one day, looking down, and asking this very question:

> Although naked apes [humans] have pubic hair, surely our hairy cousins don't? How could I test my hypothesis? I knew that there was a stuffed chimpanzee in the Grant Zoological Museum at University College London and I called in on the way to my laboratory. Alas, he was a juvenile, which left the question open. A brisk walk across Regent's Park to inspect the adult gorillas in their splendid new pavilion at London Zoo strengthened my suspicion, and this was later confirmed by a visit to the chimpanzees at Whipsnade Zoo north of London. Indeed, all the species of apes, Old World monkeys and New World monkeys seem to be less hairy in the pubic region than elsewhere; fur is present but it is short and fine.

Weiss speculates that one of the main reasons that human beings uniquely evolved a "thick bush of wiry hair" around their genital regions is to visually signal sexual maturation. (It also likely serves as a primitive odor trap and aids in the wafting of human pheromones.) So pubic hair acts as a furry advertisement, indicating for prospective sexual partners that mating with that individual could potentially be a fruitful exercise in genetic perpetuity. Weiss believes that showing off our fecundity this way suggests pubic hair would have arisen only after we became "naked apes," causing it to stand out so vividly against the backdrop of an otherwise hairless body.

Another fascinating thing about pubic hair is its unusual texture and composition compared with the rest of the hair on our bodies and heads. You can't quite use it to floss with (trust me), but pubic hair is considerably thicker than either axillary (underarm) hair or that on our legs, chests (for some, backs), and scalps. I'm probably not the only one who shudders to think of an alternative path of natural selection, one in which the hair on our heads evolved to be of pubic proportions—just consider what the average barbershop floor might look like at the end of the day. It's not entirely clear why pubic hair is so distinctly thick, short, and, usually, curly, but the biologist Anne Clark from SUNY Binghamton did point out to me (while we were hiking on the feathery and furless Kapiti Island in New Zealand, which made it all the more memorable) that anything else would be rather impractical. To have long, flowing, stylish locks growing down there wouldn't be terribly convenient, especially given the logistics of sexual intercourse.

But, as Weiss points out, although pubic hair had its signaling advantages, it also came with a cost. And this cost goes by the name of *Phthirus pubis*—more commonly known as crabs. The evolutionary story of crabs is remarkable, and it's one that Weiss relays in an issue of the *Journal of Biology*. If you've ever marveled at the similarity between human pubic hair and the coarse texture of gorilla hair (and let's face it, who hasn't), you're already on the right track:

On the basis of morphology, human *Phthirus pubis* is closely related to the gorilla louse, *Phthirus gorillae . . .* Molecular phylogeny indicates that human pubic lice diverged from gorilla lice as recently as 3.3 million years ago, whereas the chimp and human host lineage split from the gorilla lineage at least 7 million years ago. Thus, it seems clear that humans acquired pubic lice horizontally, possibly at the time of the *Phthirus* species' split and probably directly from gorillas. Because they were already adapted to the coarse body hair of the gorilla, crabs would have found a suitable niche in human pubic hair.

That's right. We got crabs from gorillas. But get your head out of the jungle gutter. Weiss speculates that our ancestors acquired these ravenous parasites not through interspecies sex but as a consequence of ancient humans butchering and eating gorillas. This close contact with gorilla carcasses would have enabled the gorilla louse (*Phthirus gorillae*) to jump hosts and mutate in accordance with the eventual evolution of human pubic hair—what must have seemed a cozy and familiar environment—to become the *Phthirus pubis* species we know and loathe today (much as bush-meat slaughter practices allowed retroviruses to invade humans from chimpanzees more recently).

Regardless of how they came to be there, crabs have unfortunately become part and parcel of our species' pubis. Intriguingly, however, recent behavioral innovations in our species' cultural evolution—particularly, modern grooming habits and the aesthetic stylizing of our pubic hair regions—have begun prying loose these pesky critters' grip on us. Some health clinics have noted a significant fall in the occurrence of pubic lice, especially among patients who shave all or some of their pubic hair. (And even if only their sexual partners shave down there, the risk of acquisition in the patients themselves should still be substantially less than for those who mate with partners whose genitals are hidden in the type of thick copse that crabs delight in.) This isn't an entirely new phenomenon. Prostitutes in medieval times would often wear "merkins" (or pubic wigs) after shaving their genitalia to help control their pubic lice.

But before you go scheduling your next full Brazilian wax, consider that pubic hair does appear to offer some degree of protection against even nastier bacterial and viral infections. Although the diagnosis of pubic lice has seemingly plummeted as a direct result of human vanity in both sexes, cases of gonorrhea and chlamydia have increased over the same period, a correlation that may not be merely coincidental. Damned if you do shave, damned if you don't.

Still, the "hairlessness norm" is gathering a lot of steam, particularly in Western countries. Several recent studies reveal just how common shaving one's nether regions has actually become. In an issue of *Sex Roles*, the Flinders University psychologists Marika Tiggemann and Suzanna Hodgson found that 76 percent of a sample of 235 female undergraduates from Australia reported having removed their pubic hair at some point in their lives. Sixty-one percent currently did so, and half of this sample said that they routinely removed all traces of their pubic hair. The current trend for men appears to be no different. In a separate study the same year, with her colleagues Yolanda Martins and Libby Churchett, Tiggemann reported in *Body Image* that of 106 gay men, 82 percent had removed their pubic hair at least once. And lest you think that this is an artifact of gay male culture, straight men weren't far behind on this measure. Out of a sample of 228 hetero-sexual men, 66 percent reported doing the same. Irrespective of sexual orientation or gender, the investi-gators discovered that the primary motivation for pubic hair depilation is related to concerns with one's appear-ance (in contrast to health-related motivations).

Let's also not forget that many individuals are put off by the idea of cunnilingus or fellatio because of those pesky pubic hairs that can lodge inadvertently in their gratifying throats. In fact, this was the theme of an episode of *Curb Your Enthusiasm*, where Larry David had to embarrass-ingly explain this bothersome tickle to a rather serious doctor. But now this is turning into a different type of story altogether.

In any event, pubic hair coiffure is not a zero-sum game. Typing "pubic hair styles" into my Google search bar yielded 467,000 hits at the time of this writing, every single one of which I was hesitant to click on—until I got home from the public library, of course.

Bite Me: The Natural History of Cannibalism

While I was strolling not so long ago through one of the dimly lit backrooms of a wing in the National Galleries of Scotland, my inner eye still tingling with thousands of Impressionist afterimages, pudgy Rubensian cherubs, and Gothic quadrangles, one irreverent painting leaped out at me in a very contemporary sort of way. It was part of an early-sixteenth-century triptych showing what appeared to be a solemn, middle-aged clergyman in gilded ecclesiastical robes commanding three naked adolescent boys before him in a bathtub.

Now, I must say, my first thought on seeing this salacious image was that the Catholic Church has been an ephebophile's haven for far longer than anyone has ever realized. But my uneasiness was put to rest once I leaned in to read the caption, which stated that the Dutch artist Gerard David, a prolific religious iconographer based in Bruges, Belgium, was merely painting a scene of starvation cannibalism. *Phew!* What a relief it was only an innocent case of anthropophagy (the eating of human flesh by humans) and nothing more sinister than that. The boys had been killed by a butcher, you see, and their carcasses were salting in a makeshift vat awaiting ingestion by famished townspeople. Fortunately, that most notorious child lover himself, Saint Nicholas—the middle-aged clergyman—just happened to be passing through town when he caught wind of the boy-eating scandal and resurrected the lads in the tub.

In any event, my time in Edinburgh offered plenty of food for thought on the subject of human meat. From the art gallery, my partner, Juan, and I galloped over to the Surgeons' Hall Museum, where we wandered through aisles packed floor to ceiling with pickled gangrenous feet, hairy severed arms of industrial-age elderly women, trephined heads, and sundry sickly genitals. Also on display was an elegant leather notebook, composed of a substance resembling cowhide but, in fact, made of the skin of the famous corpse supplier-cum-murderer William Burke.

And all of this got me thinking about the logistics of cannibalism. The slick commercialization of the food industry has changed things dramatically, but there were, at one time, relatively frequent conditions—crop failures, habitat depletion, famine—in which cannibalism would have had lifesaving adaptive utility for our species. One pair of anthropologists, for example, actually crunched the numbers, concluding that the average human adult provides sixty-six pounds of edible food, including fat, connective tissue, muscle, organs, blood, and skin. Protein-rich blood clots and marrow are said (by the rare connoisseur) to be special treats. At least one prominent evolutionary theorist, Lewis Petrinovich, has argued that cannibalism is a genuine biological adaptation common to all human beings—including those of you gripping the toilet seat as you're reading this.

Anthropophagy routinely emerges, says Petrinovich, under predictable starvation conditions, and at least during our early evolution human cannibalism was not as rare as you might think. Today the term "cannibalism" conjures up sensationalistic stories of helicopter crashes in remote montane regions in the Andes, serial killers, or failed nineteenth-century expeditions to the Arctic. But our deeper history suggests that it would not have been an entirely uncommon occurrence. "The point is that cannibalism is in the human behavioral repertoire," writes Petrinovich in *The Cannibal Within*, "and probably is exhibited for a number of reasons—a common one being severe and chronic nutritional deprivation. A behavior might be exhibited only under extreme circumstances and still be part of our biological inheritance, and the fact that its course follows a systematic pattern argues against the hypothesis that it is psychotic in character."

Petrinovich wends his way through a human history littered with the gnawed-on bones of our cannibalized ancestors, revealing that—contrary to critiques arguing that man-eating is a myth conjured up by Westerners to demonize "primitives"—we really have been gobbling each other up for a very, very long time. We're just one of thirteen hundred species for which intraspecific predation has been observed. Among primates, cannibalism can usually be accounted for by nutritional and environmental stress, or it appears as a reproductive strategy in which baboons, for example, consume unhealthy infants to make way for more viable offspring.

Pinpointing the specific factors that cause cannibalism is a rather difficult affair in the laboratory, mainly because of those pesky university ethics review boards. Still, an intrepid Japanese researcher shrugged off these consider-ations and induced cannibalism among a captive population of squirrel monkeys by feeding the pregnant females a low-protein diet. This led to a high rate of spontaneous abortion and the mothers' devouring their aborted fetuses—a much-needed bolus of protein. Now imagine doing this same study with human beings under similar controlled laboratory conditions. Rather horrific, I should say, but that doesn't mean the findings couldn't generalize to our own species. And don't get me started on the many ways that mammalian mamas feast on placental afterbirth. Some of our own prefer it with a dash of paprika, others as a spaghetti and meatballs dish.

But the fact that cannibalism in primates, including human beings, is motivated by starvation is precisely the point that Petrinovich is arguing. Where he differs from other evolutionary theorists, however, is in his assertion that anthropophagy represents a true *adaptation* in our species, just as cannibalism does for other animals. It is not simply an anomalous behavior found in a handful of depraved individuals. Such people do exist, to be sure— like this man who was so curious to know what his own flesh tasted like that he . . . well, I'll let the clinical psychiatrists who examined him tell you in their own words:

> After he cut the first toe, he first showed it to his flat-mates before he ate it raw while he walked the streets. He chewed as much of the bone as possible and then spat it out. He recalls eating it "for the experience" and that it was a "once in a lifetime opportunity to eat human flesh." He was excited by the shock value of doing so. The second toe was cooked in an oven before eating. In between cutting his toes he continued to work on renovating houses.

That man is presumably wearing special orthopedic shoes now. But again, whereas cannibalism can certainly be deviant, in other cases it's even somewhat routine. Our close cousins the Neanderthals were essentially carnivorous predators and were driven to cannibalism at the end of the last glacial maximum in the face of dwindling numbers of large game animals. Osteo-archaeological research at a cave in southeast France yielded a bundle of roasted Neanderthal bones from about six individuals, haphazardly discarded bones that had been deliberately de-fleshed and disarticulated and the marrow extracted.

As for our own species, the Aztec were notorious for their bloodthirsty sacrifice and cannibalism rituals. These were largely symbolic religious events, but some scholars have suggested that the greasy surfeit of Aztec sacrifice victims may have also been a high-energy nutritional supplement for the wealthy elite, who had first dibs on this so-called "man corn." Actually, noncannibals may be the outliers, both historically and cross-culturally speaking. Researchers have documented evidence of ritual anthro-pophagy throughout societies in Africa (Zandeland, Sierra Leone, the Belgian Congo), South America (eastern Brazil, Ecuador, western Colombia, Paraguay), the New Hebrides (Fiji, Papua New Guinea, Vanuatu, East New Guinea Highlands), and Native America. It's appeared in industri-alized societies, too, including famine-stricken China during the Great Leap Forward (1958–62) and Soviet-era Russia.

The bottom line, says Petrinovich, is that when you're hungry enough, ravenous really, and when all other food sources—including "inedible" things you'd rather not stomach such as shoes, shoelaces, pets, steering wheels, rawhide saddlebags, or frozen donkey brains—have been exhausted and expectations are sufficiently low, even the most recalcitrant moralist among us would shrug off the cannibalism taboo and savor the sweet meat of man . . . or woman, boy or girl, for that matter. It's either that or die, and between the two choices only one is biologically adaptive.

A behavior can be adaptive without being an inherited biological adaptation, of course. But because starvation occurred with such regularity in our ancestral past, and because the starving mind predictably relaxes its cannibalistic proscriptions, and because eating other people restores energy and sustains lives, and because the behavior is universal and proceeds algorithmically (we eat dead strangers first, then dead relatives, then live slaves, then live foreigners, and so on down the ladder to live kith and kin), there is reason to believe—for Petrinovich at least—that anthropophagy is an evolved behavior. The taboo against cannibalism is useful in times of health and prosperity; groups wouldn't survive very long if members were eating one another up. Yet starvation has a way of releasing the cannibal within.

In fact, some scientists have suggested that starvation cannibalism may have been so prevalent in the ancestral past that it literally changed our DNA. Modern human populations appear to contain genetic adaptations designed specifically to combat cannibalistic viruses. Typically, when a predator species consumes a prey species, there are substantive differences in immune systems between the two, with different varieties of pathogens. But the more similar the eater and the eaten, the more vulnerable the former to debilitating food-borne disease. This is because organisms can be compromised only by parasites that have adapted to the particular environment of the host species; they require a recognizable genetic substrate to thrive.

According to the microbiologist Carleton Gajdusek, who won the 1976 Nobel Prize in Physiology or Medicine for his epidemiological research on cannibalism, this is what almost certainly happened with the New Guinea Fore people in the case of kuru, a neurodegenerative disease that devastated that population in the early half of the last century. Gajdusek traced the disease to mortuary cannibalism; women and children were eating the brains of the recently deceased as part of the local funerary rites. (Brain consumption was a ritual act, but it spiked in frequency—perhaps not coincidentally—whenever pork had fallen into short supply, so human brains also infused a dose of protein.) The interesting thing is that kuru is a variant of Creutzfeldt-Jakob disease (CJD) and probably resulted, originally, from a single case of cannibalism among the Fore of a CJD-ridden brain, with kuru then evolving on its own course. In an issue of *Current Biology*, the geneticist John Brookfield speculated that over the past 500,000 years, human beings have developed increasing variation in the gene for the human prion protein. Those who are heterozygous for this gene, he points out, were protected against CJD through cannibalism. "This sustained heterozygote advantage [was possibly] created by a lifestyle of habitual cannibalism, implying a new vision of the lifestyles of our ancestors."

As we've seen, not all cases of cannibalism are due to nutritional needs. Sociopathic individuals such as Jeffrey Dahmer, Armin Meiwes, and Issei Sagawa lived in urban environments peppered with fast-food restaurants and overflowing grocery stores, yet still they dined on people. In *SuperSense*, the psychologist Bruce Hood argues that such cases reflect *essentialist* beliefs, the idea that the victims' hidden "essences" or personality attributes are acquired by physical ingestion. It's also interesting that many such cases have a sexual component. As Margaret St. Clair wrote teasingly in the foreword of *To Serve Man*:

A Cookbook for People: "There is no form of carnal knowledge so complete as that of knowing how somebody tastes." I suspect there's some truth to that uncomfortable joke. Essentialist beliefs may account for our species' peculiar history of medical cannibalism as well. The conquistadores and their New World heirs were known to have used human fat from agile natives to grease their arthritic joints. Long before Armin Meiwes was even a twinkling in his mother's eye, pregnant Aché women of Paraguay were nibbling on boiled penises in the hopes that it would bring them sons.

So with all of these scenes swimming in my head, and pragmatist that I am, I'm left wondering why, exactly, it is that the consumption of already dead human bodies is such a taboo, especially for societies in which the soul is commonly seen as fiitting off at death like an invisible helium balloon. If you subscribe to such dualistic notions, after all, the body is only an empty shell that the now-liberated spirit no longer needs. Even resurrectionists could gleefully feed the impoverished with their own flesh, lest they, God forbid, allow such a bounty of edible meat to go to rot. All those wasted commercial goods, burned down to dry, gravelly dust in crematories, squirreled away behind ornate vaults, fed extravagantly to bloated sub-terranean organisms! If you'd rather not eat meat from aged or possibly diseased dead people, and if you're worried about the dignity of the individual, it would be easy enough to breed and then factory-farm braindead or free-ranging anencephalic human beings, treating them humanely, of course, but enforcing food safety standards to control for any outbreaks.

After all, let us not forget those starving people of this world, surrounded by—as some epicures swear—the most succulent meat on the planet.

The Human Skin Condition: Acne and the Hairless Ape

Humans are pimply. It's part of what sets us apart from the rest of the animal kingdom. While it's true that some form of *acne vulgaris* affects other species—it's been found in some Mexican hairless dogs and induced experimentally in rhino mice—acne is largely an affliction of our accursed species alone. (Somewhere between 85 and 100 percent of adolescents exhibit acne—and a significant minority of adults, too.) Why is the human animal so peculiar in its tendency to form volcanic comedones, papules, pustules, nodular abscesses, and, in some severe cases, lasting scars? According to the evolutionary theorists Stephen Kellett and Paul Gilbert, we probably owe these unsavory blemishes to our having lost our apish pelts too rapidly for our own good.

Although increasingly glabrous (hairless) skin probably evolved for adaptive purposes—it may have enabled our ancestors to keep cool, for example, while traveling across the hot savanna—the sure-footed pace at which genes for depilated flesh were selected posed some cosmetic problems. Kellett and Gilbert observe that the evolution of our sebaceous glands, which were accustomed to dealing with hair-covered flesh, lagged behind this change in our appearance. As a consequence, all that oily and waxy sebum, normally committed to lubricating fur, hadn't much fur to lubricate. So the sebum started to build up and clog our pores instead. (There are many issues that a person suffering from hypertrichosis—also known as werewolf syndrome—has to worry about, but acne tends not to be one of them.) Better this evolutionary account than pimples by intelligent design, in any event. What a heartless God indeed that would wind up the clock so that our sebaceous glands might overindulge in sebum production precisely at the time in human development when we'd become most acutely aware of our appearance.

It only makes matters worse that evolution has given us another distinctly human trait, and one that makes any outbreak of acne infinitely more upsetting. I'm referring to our crippling sensitivity to other minds. Although this statement is not entirely without controversy, it seems likely, based on the available evidence, that other species do not share our fine-tuned facility at taking on the rich psychological perspective of others. If this is so, then seeing the flash of disgust, or even a more innocent curiosity, reflected in other human eyes whenever they steal away to our physical flaws triggers in us an aversive state entirely original to our species. Anyone who has ever had a ripe, loathsome pimple placed strategically upon the tip of her nose by the epidermal fates has felt this painful interpersonal state.

Consider a scene from Jean-Paul Sartre's *No Exit*, in which three strangers come to realize that they've just been cast to hell, which is, strangely enough, an average, furnished drawing room. The Devil's insidious rub, however, is that there are no windows, no mirrors, and no sleep permitted in this room. Even the characters' eyelids are paralyzed, disallowing them the simple luxury of blinking. Their exquisite little torture is for all eternity to be under one another's unrelenting glare. Inez, a sadistic lesbian, knows just how to push the buttons of the other female in the room. "What's that?" she asks, examining Estelle's face. "That nasty red spot at the bottom of your cheek. A pimple?" "A pimple?" replies the frantic, mirror-deprived, pampered debutante Estelle. "Oh, how simply foul!"

Sartre's chthonic allegory bears a striking resemblance, in fact, to the sort of living hell that many acne sufferers report experiencing on an everyday basis. For a report in the *British Journal of Health Psychology*, for instance, the psychologists Craig Murray and Katherine Rhodes interviewed around a dozen members of an online acne support group, who'd been prescribed antibiotics or hormone treatments for their condition and suffered from acne for at least a full year. "Michelle" eloquently describes what it feels like to meet someone new, face-to-face:

> I can feel the self-consciousness slowly consume me as the conversation progresses. Eventually I cannot even retain my train of thought and become tongue-tied. I unravel. I do become overwhelmed at what others might be thinking—I don't usually assume what they might be thinking with any specificity. That would be too painful an endeavour. But I do give them a generalized voice. I acknowledge to myself that they have seen the acne and most likely think less of me due to its presence.

Another woman, "Laura," notes:

> When I'm talking to people, I always stare them straight in the eye to watch if their pupils wander to other places on my face where I have a zit. And they usually do.

Obviously, acne anxiety isn't just a female problem. It's arguably even worse for some males. One sufferer, "Karl," explains why:

> Society doesn't allow [males] to wear makeup so we have to go out in the world in embarrassment. And if we tell people that we are feeling depressed or are concerned with our looks, we are looked down on as weak and pathetic, especially by other males.

Speaking of thinking about others' thoughts, I know what you're thinking: those who'd judge a book by its cover or ostracize a poor, pimpled pal in these ways ought to be scorned in public themselves. I very much agree. But in spite of our sympathy—perhaps empathy—for those suffering from such visible skin disorders, even the most kindhearted among us appear to associate acne sufferers with undesirable characteristics. At least these were the results reported by the psychologist Tracey Grandfield and her colleagues in the *Journal of Health Psychology*. Using a variation of the Implicit Association Test—an empirical measure used to get at people's unconscious attitudes and beliefs—the authors found that compared with our ratings of clear-skinned individuals, we're quick to associate unpleasant concepts (such as "brutal," "bad," "ugly," "angry," "aggressive," "vomit," and "mean") with acne sufferers. These authors reason that this unfair, unconscious, and visceral reaction to those with serious acne also betrays our evolutionary origins. Research indicates that significant disruptions of the skin surface—showing blood, pus, or flaking skin—elicit greater disgust and contamination fears among observers than "cleaner" disruptions, such as vitiligo and port-wine stains.

For many people, especially those who score high on the

personality dimension of social sensitivity, acne is not simply a nuisance; rather, it can seep ruinously into the individual's core self-concept and lead to severe mental health problems, even rivaling the distress associated with facial disfigurement from burns or accidents. One-third of New Zealand teenagers who described themselves as having "problem acne" had thoughts of suicide, one-quarter displayed clinically significant levels of depression, and one-tenth had high levels of anxiety. As long ago as 1948, the clinicians Marion Sulzberger and Sadie Zaidens concluded that "it is our considered opinion that there is no single disease that causes more psychic trauma, more maladjustments between parents and children, more general insecurity and feelings of inferiority and greater sums of psychic suffering than does acne vulgaris."

That was more than sixty years ago, and of course the acne-treatment industry has grown enormously since then. (So has the psychiatric subfield of psychodermatology.) Although not always without its own unpleasant side effects, there is an ever-flourishing pharmaceutical garden of ointments, creams, and pills today that the acne sufferers of pus-filled yore could only dream about. Still, not all such treatments are equally available to those with acne, there are considerable individual differences in response to drugs, and a fail-safe "cure" remains elusive. In fact, I suspect that by contrast to previous generations, those who experience moderate to severe acne today find themselves even more depressed than those who came before. Just as overweight people who have tried every diet without success often report feeling powerless over their condition, anyone who has attempted unsuccessfully to rid himself of acne with a wide range of treatment options may feel even more ashamed than ever.

It's little solace to these poor souls that the condition,

like most other human traits, is determined by some combination of genes and environment. How, exactly, our DNA interacts with diet, face-washing habits, exposure to the sun, or any other factor remains little understood. Yet, just as some members of that commiserating breed, the Mexican hairless dog, are more prone to acne than others, so, too, are some of us hairless apes. In balance, acne seems to have less to do with how we live than with the family we were born into. Intriguingly, and for reasons that are still unclear, certain human populations, such as the Kitavan Islanders of Papua New Guinea and the Aché of Paraguay, are spared the blackhead plague. Although their diets and lifestyles are very different from our own, so are their genes.

Yes, less is more in the present case. But few of us are so lucky as to have the silken pelage of a Wookiee or find ourselves born an indigenous Kitavan Islander, and the lifelong zitless are extremely rare. The best-case scenario is that your skin isn't too much of a workaholic when it comes to sebum production, and so, like everyone else, you'll get only the occasional breakout here and there. Ideally, in terms of your psychological health, the pimples will be hidden somewhere over *there*, rather than *here* on that blinking marquee that is your face, unprotected from the elements.

Whether your acne disappears by your teens or not until your forties, your sebaceous glands will one day, I promise you, run dry as an ancient riverbed. Although you could have easily gotten lost in her glorious wrinkles, for example, I don't remember a single zit on my eighty-nine-year-old grandmother's face when that non-ethereal husk of hers was peaceably rehydrated by formaldehyde. So remember, all of you with reddened hides in hiding, those in sore, oozing discontent, acne is a passing cosmetic calamity. There's no shame in shame, so ask for help if you need it. You aren't alone in your distress, but save some worrying for those slowly gestating, well-earned wrinkles to come. Above all, be kind to your inner ape that lost its fur in haste.

PART III

Minds in the Gutter

Naughty by Nature: When Brain Damage Makes People Very, Very Randy

If you're reading this, my guess is that you're a materialist holding the logical belief that the human brain—with all of its buzzing neural intricacies, its pulpy, electrified arabesque chambers and labyrinthine coves—has been carved out over countless eons by the slow-and-steady hand of natural selection. You will grant, then, that specific brain regions evolved because they generated behaviors that were beneficial to our ancestors. When one part of the brain is compromised—through injury, disease, or some other unfortunate event—the constellation of symptoms that result are often remarkably specific. "The brain is the physical manifestation of the personality and sense of self," writes the neuroscientist Shelley Batts in *Behavioral Sciences and the Law*, "and focal damage to brain areas can result in focal changes in behavior and personality while leaving other aspects of the self unchanged."

Not to get too technical, but if you're unlucky enough to develop a lesion that interferes with the functioning of your dorsolateral prefrontal cortex—a specialized patch of neural tissue that's intricately braided into your anterior cingulate cortex—then your working memory, strategy-formation, and planning skills are going to take a major nosedive. Suddenly something as simple as coming up with a list of groceries becomes a major achievement.

Most of us have sympathy aplenty for those patients whose brain disturbances have interfered with their everyday cognitive abilities. We're perfectly willing to accommodate their intellectual disabilities by helping them create a new mnemonic strategy or giving them a pat on the back or a word of encouragement when they're trying to remember someone's name (because, frankly, who hasn't struggled with these things?). Yet when chunks of gray matter that have evolved to control and inhibit, say, our sexual appetites and other bacchanalian drives experience a similar catastrophic blowout, are we so understanding? What if those impairments lead their victims to display ... oh, I don't know, let's call them *moral disabilities*? Cases of libidinal brain systems going haywire have our kindhearted, humanistic materialism rubbing elbows—or butting heads—with our belief in free will and moral culpability.

Although Klüver-Bucy syndrome is relatively rare, it's one of the most notorious neurological causes of a complete breakdown in one's ability to control sexual urges. In 1939, the neuroanatomists Heinrich Klüver and Paul Bucy removed the greater portions of both temporal lobes and the rhinencephalon from the brains of rhesus monkeys. Initially, these scientists were interested in studying how mescaline administration produced seizures similar to temporal-lobe fits in epileptic patients and so were attempting to isolate the effects of those drug-addled brain regions. Among a host of other peculiar effects of this rather cruel vivisection, however, the monkeys became incredibly randy, displaying a prominent and indiscriminate desire to copulate. The first documented case of full-blown Klüver-Bucy in humans arrived in 1955, when an epilepsy patient underwent a bilateral temporal lobectomy (a surgical excision of the lobes) and subsequently developed a ravenous sexual appetite, among other things. More often, the syndrome appears in lesser degrees, precipitated by a nasty insult to the medial temporal lobe. It might result from a case of herpes encephalitis or Pick's disease, or from trauma and oxygen deprivation. Not all such patients experience hypersexuality, mind you, but some do. Other symptoms aren't terribly appealing either, however; they include hyper-orality (a compulsive desire to put things in one's mouth), apathy, emotional unresponsiveness, and various sensory disorders.

Dramatic case studies illustrating the devastating effects of Klüver-Bucy syndrome abound in the clinical literature, and they raise intriguing philosophical questions for us to consider with respect to the sheer physicality of "free will." That some patients so stricken are overcome with excessive carnal urges and are not simply using the disorder as a convenient excuse to become freely promiscuous, lewd, and lascivious is perhaps best demonstrated by a 1998 study by the neurologist Sunil Pradhan and his colleagues. In this report, a group of boys between the ages of two and a half and six began to exhibit hypersexualized behaviors after partially recovering from comas induced by herpes encephalitis. One to three months after emerging from the comatose state, "all seven children," note the authors, "demonstrated abnormal sexual behavior in the form of rhythmic hip movements [sexual thrusting] (two patients), rubbing genitals over the bed (two patients) and excessive manipulation of genitals (all seven patients)." Were these children just helpless, hapless puppets of their ancient, pleasure-driven brains? The authors believe so: "As all patients [were extremely young], with no possibility of environmental learning of sex, these movements most probably represented phylogenetically primitive reflex activities."

It may be awkward enough telling other parents why your preschooler is humping everything in sight—just try rehashing the foregoing description of Klüver-Bucy syndrome to your friends at the day-care center—but we do tend, as adults, to be mostly forgiving of a child's improprieties. When this sort of hypersexuality strikes a postpubescent individual whose sexuality is driven by orgasm-propelled desires, things become more interesting—at least, again, in a philosophical sense. Although it would be entirely inaccurate to portray Klüver-Bucy patients as sex-crazed lunatics, they very often display behaviors that would be considered inappropriate by conventional standards. One gentleman in his early seventies, for instance, hugged a female parishioner at his church and repeatedly kissed her. According to the clinical case report, he then asked the shocked woman, "Why don't we do it again?" Over the ensuing years, his sexual fantasies skyrocketed, and his hyperorality became unmanageable. The report notes that, according to his wife, "he would put any object in his mouth, including dog food, candles, adhesive bandages, and his wedding ring. His appetite seemed insatiable . . . He died at age 77 years of asphyxiation on several adhesive bandages."

In a letter to the editor of *European Psychiatry*, two physicians describe the case of a fourteen-year-old schoolgirl ("Ms. A") who, prior to developing Klüver-Bucy syndrome after being in an encephalitis-caused coma, "was an intelligent and social girl with a good academic record." This quiet, well-behaved teenager became somewhat challenging, to say the least, after recovering from her illness. You think *you're* raising a difficult teen? Consider what these parents were dealing with:

> The patient started . . . disrobing in front of others, manipulating her genitals, and making sexual advances toward her father. She would lick any object lying on the ground and whenever she got an opportunity, she would rush to the toilet and try to put urine and feces into her mouth [urophagia and coprophagia, respectively].

In another case, an epileptic woman underwent an unsuccessful left-temporal lobectomy to help stop debilitating seizures. Klüver-Bucy symptoms, including hypersexuality, emerged following the surgery. She began masturbating in public and aggressively soliciting her family members and neighbors for sex. After having another seizure, she was brought to the emergency room, where, after half an hour in the waiting area, she began performing fellatio on an elderly cardiac patient. (This may or may not be one of the few examples where one person's syndrome is another's lucky day; it's also unclear if this was a display of hypersexuality or hyperorality, but it's inevitable, perhaps, that the twain should occasionally meet.)

Other temporal-lobe epileptics have also exhibited hypersexuality in the "postictal" state, which is the period of recovery time following a seizure. The neurologist Vanessa Arnedo and her colleagues presented a case of a thirty-nine-year-old man who began having semi-frequent seizures during the middle of the night. After nocturnal convulsions, he'd sleep for another ten minutes, wake up, and then rape his wife. (In the authors' more delicate wording, he was described as "becoming sexually

aggressive toward his wife by forcing intercourse.")
Importantly, however, "the tremendous remorse and
abhorrence for what he had done when he learned of his
actions led him to pursue possible surgery mainly to
eliminate this postictal behavior." Other people with
similar epileptic profiles also become hypersexualized in
the postictal state. To his later horror, one man motioned
for his twelve-year-old daughter to join him and his wife in
the bedroom following a nighttime seizure.

It is in these last few examples, where Klüver-Bucy
syndrome manifests itself in criminal behavior such as rape
or child molestation, that our materialistic convictions are
really put to the test. In 2003, the neurologists Jeffrey
Burns and Russell Swerdlow described how an otherwise
well-behaved forty-year-old man developed a case of
"new-onset pedophilia" after suffering the appearance
of a right-orbitofrontal tumor. The man denied any pre-
existing interest in children; he did have a predilection for
pornography before the tumor, say Burns and Swerdlow,
but now he was downloading child porn and making
subtle sexual advances to his prepubescent stepdaughter.
His hypersexuality applied to full-grown women, too—so
much so, in fact, that he couldn't keep himself from
fondling female nurses and staff during a neurological
examination. Long story short, when the man's tumor was
removed, his prurient interests and behaviors all but disappeared, and since he was no longer deemed a threat to
his stepdaughter, he returned home. But his headaches
returned, his tumor regrew, and so did the criminal
impulse. A "re-resection" of the tumor was accomplished,
the man became a good citizen again, and, as far as we
know, that remains true today. In a more recent case co-
published by the famed neuroscientist Oliver Sacks, a
fifty-one-year-old without any criminal history had part of

his right temporal lobe removed to prevent seizures. Following this, he developed telltale signs of Klüver-Bucy, including hypersexuality. His was another case of "new-onset pedophilia," but, as Sacks laments, in spite of this he was nevertheless sentenced to several years in prison for downloading child porn.

What's the takeaway message? I'll let you do the hard work of thinking through the implications for our belief in free will and how it might or might not apply to Klüver-Bucy syndrome. But another intriguing question emerges, too: If an otherwise "good" person's brain can be rendered suddenly morally disabled by an invasive tumor or an epileptic short circuit, subsequently causing him or her to do very bad deeds, then isn't it rather hypocritical to assume that a "bad" person without brain injury—whose brain and neural functioning are organized by the complex interplay between genes and experience (and every single phenomenal aspect of whose mind is therefore physically constrained)—has any more free will than the neuro-clinical case? After all, people have zero control over the particularly idiosyncratic brain they're born with, and very little control over their early life experiences, which, in turn, can only work with whatever congenital neural substrate is already there.

Perhaps it's just a matter of timing: the "good" are born with brains that can "go bad," whereas the "bad" are hog-tied by a prospective morally disabled neural architecture from the very start. And although it may be less common, if a "bad" person behaves in an upstanding manner, could that be the result of fortuitous brain damage or epilepsy, too? Should we not regard such a person highly if he saves a child from a burning building because, like the man gesturing to his twelve-year-old stepdaughter to join him for sex, that isn't really him?

At issue is not holding healthy people to a "higher standard" or making excuses for criminals, but instead simply recognizing that the degree by which we have control over our actions—any of us—is *entirely* neurologically based. Free will is physical. And if indeed it's all brain-based in the end, a position you probably subscribed to at the start of this essay, this also includes the extent, the sophistication, and the parameters by which one can even objectively contemplate free will (such thinking is constrained by brain-based cognitive capacities, after all). The shocking truth is that we're only as free as our genes are pliable in the slosh of our developmental milieus.

How the Brain Got Its Buttocks:
Medieval Mischief in Neuroanatomy

There are so many specializations within the brain sciences that even the sharpest brain has scarcely enough brain-power to learn everything there is to know about itself. But if there's one fact that the teacup-Yorkie-sized prune in your head might want to ponder, it's that it shares a peculiar past with something considerably lower in your anatomy—your genitalia. I don't mean that our brains and reproductive organs share some embryological or evolutionary history, but rather that they were once (and, to some extent, still are) entwined in the *language* of the body. What this odd story reveals is that the ancient anatomists were major dickheads. We all were, back then. According to ancient nomenclature, even women had penises in their brains.

Régis Olry, an anatomy professor, and Duane Haines, a neurobiologist, brought the whole sordid tale to light in an intriguing pair of articles for the *Journal of the History of the Neurosciences*. These historians of neuroanatomy (yes, there is such a profession, and we should all be grateful for it) reviewed a very old, circuitous medical literature and found that the human brain was once described as comprising its very own vulva, penis, testicles, buttocks, and even anus. Not surprisingly, it was men doing all the classifying and labeling. In fact, part of the cerebrum is still named in honor of long-forgotten prostitutes—which I'll get to in a moment.

In their first article over ten years ago, an epoch in academic terms, Olry and Haines revealed the surprising origins of the term *fornix*. For those illiterate in neuro-anatomy, the fornix is an arching band of nerve fibers that connects the hippocampus and the limbic system and spans certain fluid-filled chambers of the brain known as ventricles. You'd have numerous and noticeable problems if your fornix weren't functioning properly, including serious impairments in spatial learning and overall navigation.

Some basics of etymology. Although today *fornix* is reserved almost exclusively for anatomical structures—there's also a fornix of the conjunctivae, which connects the membranes of the eye, as well as several other bodily fornices, but let's move on—the word originally held an architectural connotation, coming from the Latin for "arch." Olry and Haines point out that Roman architects during the first century B.C. created wooden rooms with vaulted ceilings, called *fornices*. When such rooms were made of brick, they were called *camerae* (there's a separate etymological history involving the modern-day camera and these brick-arched Roman rooms, but we're focusing on the fornix here).

Now, none of this is terribly salacious, and it's quite possible that the first neuroanatomist ever to use this term, the seventeenth-century Englishman Thomas Willis, had nary a dirty thought in mind. But it's also a fact that the wood-vaulted rooms of old were used expressly for the plying of a particular trade in ancient Rome, prostitution (hence *fornication*). "The real etymology of the term 'fornix,'" concluded Olry and Haines, "is therefore related to the form of the roof of the third ventricle, but also to the sexual intercourse which occurred in such rooms, these rooms being compared with this ventricle." It's merely an ironic twist that the fornix helps to regulate human sexual behavior as part of the limbic system; as the authors point out, the name was bestowed long before anyone knew this function.

In any event, once they'd put the fornix to bed, Olry and Haines waited another decade or so before they revisited the sexy third ventricle. In a follow-up article, they exposed some more, rather curiously named features from the same part of the brain. When the mid-sixteenth-century Italian anatomist Matteo Realdo Colombo peered into the small recess adjoining the anterior commissure and the dividing line of the fornix's two columns, report Olry and Haines, he saw what looked like a lubricated vulva—and called it the *vulva cerebri*. Perhaps that's not too surprising, given that Colombo is also widely credited as being the anatomist who first "discovered" the clitoris (the real one).

The authors point out there's a bit of a mystery about precisely which hole Colombo was poking with his Italian probe. It might, in fact, have been the more posterior opening identified by the seventeenth-century Dutch anatomist Isbrand van Diemerbroeck, who found, in Colombo's groove, "the hole of the anus." Your brain's anus, incidentally, is what we'd now call the "common posterior opening" of the midbrain's aqueduct, which spills into the third ventricle. There are so many defecation-related puns about intelligence to be made here that my mind is cramping up, so, shit, I'll just leave that part up to you assholes.

Now, van Diemerbroeck didn't just see lady bits in the

brain; if anything, he and his fellow anatomists envisioned it as an essentially hermaphroditic organ. After all, not only did it have a *vulva cerebri*; it also possessed a distinctive *penis cerebri*. René Descartes may have celebrated the pineal gland as the "seat of the soul," but for the less metaphysically minded van Diemerbroeck, as well as one of Descartes's contemporaries, the Danish physician Thomas Bartholin, that structure was more like a penis. This metaphor may have its roots, explain Olry and Haines, in the position of the gland above and between the brain's colliculi, which had already been compared with testicles.

This cockeyed term, *penis cerebri*, proved too embarrassing for future scholars and quickly shrank into disuse. Yesterday's penis is today's soulless pineal gland (a stiffer term, to be sure). By the mid-eighteenth century in France, a real buzzkill of an anatomist by the name of Jacques-Bénigne Winslow was already looking back in disgust at his forebears' indelicate classifications; the ancients, he thought, had their heads in the gutter when it came to what was in their heads. Winslow held these founding fathers of the neurosciences in particular disrepute for their having seen buttocks (*eminentiae natiformes*) and testicles (*eminentiae testiformes*) in the colliculi: "The names that were given to these tubercles are very impertinent, and have no resemblance with the things they were derived from." Others begged to differ, and scholars continued to refer to the buttocks and testicles in our brains for centuries after Winslow huffed and puffed about the matter, even into the twentieth century. Eventually, however, academic prudery eclipsed asinine antiquarianism.

Still, a spunky remnant of those lost days of brainiac debauchery did slip into the present-day vocabulary.

According to Olry and Haines, the glandular portion of the pineal gland can be traced back to its bulbous terminological predecessor, the glans penis. Today we know that the pineal gland produces melatonin, a chemical central to regulating your sleep-wake cycle. So the next time you have jet lag, blame it on your penis. And if I haven't nursed the history of the mammillary bodies—those small round bodies on the undersurface of the brain that are believed responsible for adding smell to recognition memory—that's only because it's too easy.

Olry and Haines weren't the first to scratch their heads over this lurid labeling of neuroanatomical regions. Joining the prudish Winslow in his disdain, the French anatomist Joseph Auguste Aristide Fort observed in 1902 that the anatomists of past centuries "enjoyed giving indecent names to the different parts surrounding the third ventricle." But Olry and Haines revealed exactly how these medieval anatomists cast their libidinous eyes upon the gray matter and saw not only the glistening engine of our thoughts but also our private parts.

Lascivious Zombies: Sex, Sleepwalking, Nocturnal Genitals—and You

It may seem to you that much like their barnyard animal namesake, men's reproductive organs the world over participate in a mindless synchrony of stiffened salutes to the rising sun. In fact, however, such "morning wood" is an autonomic leftover from a series of nocturnal penile tumescence (NPT) episodes that occur like clockwork during the night for all healthy human males—most frequently in the dream-filled rapid eye movement (REM) periods of sleep from which we're so often rudely awakened in the a.m. by buzzers, mothers, or others.

For those with penises, you may be surprised to learn how frequently your member stands up while the rest of your body is rendered catatonic by the muscular paralysis that keeps you from acting out your dreams. (And thank goodness for that. Carlos Schenck and his colleagues from the Minnesota Regional Sleep Disorders Center describe the case of a nineteen-year-old with sleep-related dissociative disorder crawling around his house on all fours, growling, and chewing on a piece of bacon—he was "dreaming" of being a jungle cat and pouncing on a slab of raw meat held by a female zookeeper.) Scientists have determined that the average thirteen- to seventy-nine-year-old penis is erect for about ninety minutes each night, or 20 percent of overall sleep time. With your brain cycling between the four sleep stages, your "sleep-related erections" appear at eighty-five-minute intervals lasting, on average, twenty-five minutes. (It's true; they used a stopwatch.) As far as I'm aware, there aren't many well-developed evolutionary theories or a proposed "adaptive function" of NPT, but we do know that it's not related to daytime sexual activity, it declines (no pun intended) with age, and it's correlated positively with testosterone levels. Although far fewer studies have examined women's nocturnal genitals, females similarly exhibit vaginal lubrication during their REM-sleep, presumably with many dreaming of erect penises.

Now, you may not think that such tedious biological details would be fodder for a moral quandary, but you underestimate our species' massive confusion when it comes to understanding how our coveted free will articulates with our genitalia. Consider the case of a young Frenchman whose sleep-related erection was interpreted by another man as a sign of sexual interest but, swore the former, was nothing of the kind. As described by a group of investigators at the annual meeting of the French Sleep Research Society in 2001, the twenty-four-year-old heterosexual male awoke to his horror with painful anal lesions. Although, having been heavily intoxicated at the time, he had no conscious recollection of any such incident occurring, this led him to deduce that he must have been raped during the night. "The legal medical examination indeed reported on visibly recent tears of the anal margin," confirmed the researchers.

Then comes the sobering whodunit. Especially disquieting was that the man's boss had slept over the night before.

The two had earlier been lounging in the pool and roasting together in the sauna. There was absolutely no evidence of date rape drugs, but alcohol, as it so often does in the south of France, flowed with relatively gay abandon that evening, and so the straight employee, being a gentleman, had invited his employer to sleep it off on his sofa while he retired to the mezzanine. Apparently, however, it was the employee who slept particularly hard that night, not the inebriated boss. The older man readily admitted that of course they'd had sex overnight, and he could only assume that his colleague's erection, combined with the fact that the other didn't resist as he mounted him, suggested that he was a consensual partner. (You thought you were a deep sleeper; imagine the somnambulistic fortitude required to snooze through your first anal penetration.) While the courts tried to sort it all out, the alleged rapist was imprisoned for two years, until finally a judge decided that both men were more or less right and the accused should be set free.

This is but one of many curious examples of sex and law intertwining. The related phenomenon of "sexsomnia" (sleep sex) has witnessed periodic public interest through a spate of high-profile cases, stories that have in turn motivated intriguing academic research on this little-known subject. Even Alfred Kinsey, the grand archivist of carnal facts, while devoting much of his discussions to the subject of wet dreams and nocturnal emissions in both sexes, didn't mention how some people act out sexually during their sleep.

Unlike the aforementioned case of the sleeping employee being the passive, immobilized recipient in unwanted intercourse, it's the sleeper who instigates the trouble in bouts of sexsomnia. Although researchers don't yet have an exact figure on the frequency of this parasomnia, most specialists believe that it's probably fairly common. Nearly all people who exhibit recurrent sexual acts while sleeping have a history of sleepwalking. In fact, many experts believe that sexsomnia is simply a variant of sleepwalking, which affects 1 to 2 percent of adults, and this is how it's presently classified in the main diagnostic manual, *The International Classification of Sleep Disorders, Revised*. Most people do not seek out clinical treatment because of either their ignorance of the condition or their embarrassment, and oftentimes their sexual "automatisms" are innocuous enough—such as fugue-state masturbation, weak pelvic thrusts, or steamy pillow talk. (More on the concept of automatism in a moment.)

In a 2007 issue of *Brain Research Reviews*, however, the psychobiologist Monica Andersen and her coauthors investigated all case studies that had, at that point, been published in the literature, and they attempted to piece together some common denominators underlying sexsomnia. They found that the most common precipitating factors of sleep sex are sleep deprivation, stress, alcohol or drug consumption, excessive fatigue, and physical over-activity in the evening. Being male and under the age of thirty-five is also a major factor, they reported; furthermore, when women do lapse into this altered nocturnal state, their actions tend to be comparatively inoffensive, moaning and masturbating rather than, like male sexsomniacs, fondling and grinding whatever is unfortunate enough to be in the vicinity of their bed that night.

One of the most extraordinary things about sexsomnia is that the sleeping person's inappropriate behaviors are sometimes directed at people that, during their waking lives, are not particularly arousing to them. In a 1996 issue of *Medicine, Science, and the Law*, the psychiatrist Peter

Fenwick describes the case of an allegedly heterosexual male cadet who was court-martialed for homosexual assault after he'd crawled into bed with another soldier and caressed that private's privates. The case was dismissed after the court accepted that the absence of an erection in the accused—sexsomnia may or may not involve erections—meant it was unlikely that the episode was "purposeful," but instead just a bizarre sleepwalking incident. Another example of atypical homosexuality in sexsomnia involved a sixteen-year-old who walked into his aunt and uncle's bedroom one night and began molesting his adult uncle.

Erections, as I hinted at earlier, complicate matters for the judicial system. One notorious case garnering international media attention, and as reviewed in *Current Psychiatry* by a group of sleep researchers from the Cleveland Clinic, centered on a thirty-year-old landscaper named Jan Luedecke, who drank far too much at a wild croquet party in the Toronto suburbs one night back in 2003 and fell asleep on a couch. "Some time later," explain the authors, "he approached a woman who was sleeping on an adjacent couch, put on a condom, and began sexual intercourse with her." From her terrified perspective, the woman awoke to discover that her underwear had been removed and a glassy-eyed Luedecke was trying to rape her. She pushed him off, ran to the washroom, and returned to find him standing there bewildered. Luedecke, who had an established history of sleepwalking behaviors, was acquitted after the psychiatrist Colin Shapiro testified for the defense that the accused was in a dissociative state when the incident occurred and therefore he was not consciously aware of his actions.

Difficult legal cases such as these hinge entirely on the demonstrability (or at least strong probability) of an automatism—a crime committed during sleep. This is a concept for which Fenwick provided one of the clearest definitions:

> An automatism is an involuntary piece of behavior over which an individual has no control. The behavior is usually inappropriate to the circumstances, and may be out of character for the individual. It can be complex, coordinated and apparently purposeful and directed, though lacking in judgment. Afterwards the individual may have no recollection or only a partial and confused memory for his actions.

In other words, sexsomniacs are basically lascivious zombies. There's presently no way to determine with absolute certainty if the phenomenon, when invoked as a defense, was really the cause or just a convenient alibi. Still, certain criteria (detailed sleep-pattern data from a nocturnal polysomnography; sleepwalking and sleep-related sex in the past; known trigger factors, such as intoxication, fatigue, and stress; timeline of the alleged assault, since episodes typically occur within two hours of sleep onset during non-REM sleep; amnesia regarding the event; no attempt to conceal or "cover up" the incident, but instead confusion) can at least aid a jury in its decision making. It's tempting, to say the least, to be skeptical that a sleepwalker could act so purposefully as to fiddle successfully with a condom wrapper yet be as conscious as an orthopteran, but the London sleep researcher Irshaad Ebrahim reminds us that sleepwalking behaviors are highly variable and can be very detail oriented, citing people preparing meals and eating, driving motorbikes and cars, even riding horses, all while getting a good night's sleep.

For those for whom sexsomnia has become a serious problem, in a legalistic sense or otherwise, the good news is that it responds well to pharmaceutical intervention. Just a small dose of benzodiazepines—most notably clonazepam—before bedtime seems to do the trick for most. You might want to consider discussing this with your doctor if you've shown a history of sexual violence during sleep or, say, you're a frequent sleepwalker and there are children in the home. (Several cases have, in fact, involved very unsettling child abuse charges being filed against alleged sexsomniacs.) But sexsomnia can be a problem even for those who live and sleep alone. After five years of waking up several nights a week with ejaculate mysteriously between his fingers, one twenty-seven-year-old was distressed to realize that he was a somnambulistic masturbator. The poor man broke two fingers when his nocturnal alter ego tore off the restraints he'd used to avoid moving in bed.

There are also those, I should point out, whose sex lives have actually benefited, courtesy of their partner's sexsomnia. Schenck and his coauthors review several such cases, including a woman who "reported infrequent and hurried sex with her [awake] husband, whom she described as distant and reluctant during wakefulness." This lady found that, with him at least, "nocturnal sex was more satisfactory, even if associated with bruises at times."

So, in closing, how do you determine if your partner's overnight poking is thoughtless or thoughtful? I'll spare you the details, but this is the very question that prompted me, several nights ago, to write this essay. Apparently, snoring during sexual behavior is a good sign and something that the partners of many sexsomniacs mention as occurring, quite out of the blue, during even the most complicated sex acts. It occurred to me also that zombified nocturnal penile tumescence episodes may be distinguished from actual conscious sexual arousal by the presence or absence of, oh, what to call it, "penile flicking." (That's not a technical term, but since I dredged the depths of the literature in vain trying to find the proper term for this voluntary up-and-down movement of the erect penis through the clenching of the pubococcygeus muscle—oh, c'mon, don't pretend you don't know what I'm talking about—please permit me a little poetic license.) I always thought such penile flicking responses must serve some communicative signaling function in our species, but apparently nobody has thought to study them from an adaptive perspective. Imagine that.

Anyway, could a sexsomniac use his social cognition to deliberately communicate a message of sexual interest by flicking his penis at his partner? It's probably not a fail-safe clue, but I suspect not. And bear that helpful hint in mind for whenever the apocalypse arrives, since God only knows it will come with its share of sex-crazed male zombies—a lot of randy gay ones too, according to many Christian conservatives.

Humans Are Special and Unique:
We Masturbate. A Lot

There must be something in the water in Lanesboro, Minnesota, because the night I stayed over there, en route to a conference, I dreamed of an encounter with a very muscular African American centaur, an orgiastic experience with drunken members of the opposite sex, and (as if that weren't enough) then being asked by my hostess to don a white wedding dress for my upcoming keynote presentation. "Does it make me look too feminine?" "Not at all," she assured me, "it's a man's dress."

Now, Freud might raise his eyebrows at such a lurid dreamscape, but if these images represent my repressed sexual yearnings, then there's a side of me that I apparently have yet to discover. I doubt that this is the case. Dreams with erotic undertones are like most other dreams during REM sleep—runaway trains with a conductor who is helpless to do anything about the surrealistic directions they take. Rather, if you *really* want to know about a person's hidden sexual desires, then find out what's in his mind's eye during the deepest throes of masturbation.

This conjuring ability to create fantasy scenes in our heads that literally bring us to orgasm when conveniently paired with our dexterous appendages is an evolutionary magic trick. It requires a cognitive capacity called mental representation (an internal "re-presentation" of a previously experienced image or some other sensory input) that many evolutionary theorists believe is a relatively recent hominid innovation. When it comes to sex, we put this capacity to very good—or at least very frequent—use. In a classic, pre-Internet-porn study (I'll get to Internet porn in a moment) by the evolutionary biologists Robin Baker and Mark Bellis, male university students were found to masturbate to ejaculation about every seventy-two hours, and "on the majority of occasions, their last masturbation is within 48 hours of their next in-pair copulation." If they're not having intercourse every day, that is to say, men tend to pleasure themselves to completion no more than two days after last having actual sex.

Baker and Bellis's quite logical argument for this seemingly counterintuitive state of affairs (after all, shouldn't men try to stock up as much sperm as possible in their testes rather than spill their seeds so wastefully in a rather infertile swath of toilet paper or a sock?) is that because there is a "shelf life" for sperm cells—they remain viable for only five to seven days after production—and because adult human males manufacture a whopping three million sperm per day, masturbation is an evolved strategy for shedding old sperm while making room for new, fitter sperm. It's a question of quality over quantity. Here are the adaptive logistics, according to the scientists:

> The advantage to the male could be that the younger sperm are more acceptable to the female and/or are better able to reach a secure position in the female tract. Moreover, once retained in the female tract, younger sperm could be more fertile in the absence of sperm competition [sexually monogamous relationships]

and/or more competitive in the presence of sperm competition [in which the woman is having sex with other men]. Finally, if younger sperm live longer in the female tract, any enhanced fertility and competitiveness would also last longer.

Unconvinced? Well, Baker and Bellis are intelligent empiricists. They also have stomachs of steel. One way that they tested their hypotheses was to ask more than thirty brave heterosexual couples to provide them with some rather concrete samples of their sex lives: the vaginal "flowbacks" from their postcoital couplings, in which some portion of the male's ejaculate is spontaneously rejected by the woman's body. As Baker and Bellis explain, "The flowback emerges 5–120 min after copulation as a relatively discrete event over a period of 1–2 min in the form of three to eight white globules. With practice, females can recognize the sensation of the beginning of flowback and can collect the material by squatting over a 250 ml glass beaker. [And here comes a useful tip, ladies . . .] Once the flowback is nearly ready to emerge, it can be hastened by, for example, coughing."

As the authors predicted, the number of sperm in the girlfriends' flowbacks increased significantly the longer it had been since the boyfriend's last masturbation—even after the researchers controlled for the relative volume of seminal fluid emission as a function of time since last ejaculation (the longer it had been, the more ejaculate, on average, was present). If only the parents of teenage boys had these findings available for the first hundred thousand years of our history, think of all the anxiety, guilt, and shame that might never have been.

In fact, even G. Stanley Hall, the father of adolescent psychology research, had a particularly nasty thorn in his paw when it came to the subject of masturbation. Hall accepted that spontaneous nocturnal emissions (that is, wet dreams) in adolescent boys were "natural," but he viewed masturbation as a "scourge of the human race . . . destructive of that perhaps most important thing in the world, the potency of good heredity." In Hall's view, the offspring of teenage masturbators would show signs of "persistent infantilism or overripeness." Boys will be boys, Dr. Hall, and—though there's lamentably no data on this—I'd still bet that those teenagers who deny themselves this natural behavior tend to have more issues than those who don't.

Now back to masturbation fantasies and cognition, and this is where it gets really interesting. Baker and Bellis's theory may be peculiarly true for human beings, because from all appearances, under natural conditions, we are the only primate species that seems to have taken these seminal shedding benefits into its own hands. Unfortunately, there has been a paltry number of studies tracking the masturbatory behaviors of nonhuman primates. Although some relevant data is probably buried in some mountain of field notes, I didn't come across any targeted studies on the subject in wild chimpanzees, and even the prolific Jane Goodall doesn't seem to have ever gone there. Nevertheless, by all available accounts, and by contrast with human beings, masturbation to completion is an exceedingly rare phenomenon in other species with capable hands very much like our own. As anybody who has ever been to the zoo knows, there's no question that other primates play with their genitalia (bonobos are notorious for this); the point is that these diddling episodes so seldom lead to an intentional orgasm.

There's not much out there in the way of proper research on ape masturbation, but some studies, here and there, do seem to document the infrequency of masturbation in other primates. In the early 1980s, scientists observed the sexual behaviors of several groups of wild gray-cheeked mangabeys for over twenty-two months in the Kibale Forest of western Uganda. There was plenty of sex, particularly during the females' peak swellings. But the researchers came across only two incidents of male masturbation leading to ejaculation. Yes, that's right. Whereas healthy human males can't seem to go without masturbating for longer than seventy-two hours, two measly cases of masturbating mangabeys were observed over a nearly two-year period.

The anthropologist E. D. Starin didn't have much luck spying incidents of masturbation in red colobus monkeys in Gambia, either. In a brief 2004 article published in *Folia Primatologica*, Starin reports that over a five-and-a-half-year period of accumulated observations totaling more than ninety-five hundred hours, she saw only five—count 'em, *five*—incidents of her population of five male colobus monkeys masturbating to ejaculation, and these rare incidents occurred only when nearby sexually receptive females were exhibiting loud courtship displays and copulations with other males.

Intriguingly, Starin says that although females weren't in the immediate vicinity, it is possible that they could still be seen or heard by the masturbating male while the incident at hand occurred. (In other words, no mental represent-ation required.) In fact, the author's descriptions of these events strike me as producing accidental, rather than deliberate, ejaculations. Not that they weren't happy accidents, but still. "During each observation," Starin writes, "the male sat and rubbed, stretched, and scratched his penis until it became erect, after which additional rubbing produced ejaculate." Also, out of the fourteen female colobus monkeys tracked during this time span, "three different females were observed possibly masturbating" by self-stimulating their genitals—only *possibly* because none of these episodes culminated in the telltale signs of colobus orgasm: muscle contractions or facial expressions or wild screams of merciless bliss.

Perhaps the most colorful report of nonhuman primate masturbation—or rather the astonishing lack thereof, even in subordinate males that aren't getting any—comes from a 1914 *Journal of Animal Behavior* study by a peculiar character named Gilbert Van Tassel Hamilton. Hamilton apparently ran something of a monkey research center-cum-sanctuary on the lush grounds of his Montecito, California, estate. He was also, clearly, a pioneering sexologist, or at least had especially liberal attitudes for his time, defending the naturalness of homosexual behavior in the animal kingdom, among other things. In justifying his research, which meant getting up close and personal with his monkeys' genitals, Hamilton opines: "The possibility that the types of sexual behavior to which the term 'perverted' is usually applied may be of normal manifestation and biologically appropriate somewhere in the phyletic scale has not been sufficiently explored."

In fact, he seems to have expected to find rampant masturbation in his animals, but to his surprise only one male (named Jocko) ever partook in such manual pleasures. "Of all my male monkeys," wrote Hamilton,

only Jocko has been observed to masturbate. After a few days confinement he would masturbate and eat part of his semen. I have reason to believe that he lived under unnatural conditions for many years before I acquired him. In view of this fact that not one of seven sexually mature monkeys masturbated after several weeks of

isolation under conditions that favored a fairly healthy
mental and physical life (close proximity to other
monkeys, large cage, warm climate), I am inclined to
believe that masturbation is not of normal occurrence
among monkeys.

Granted, Hamilton seems to have been just a tad
eccentric. Earlier in the article he reports that one of his
female monkeys named Maud liked to be mounted (and
entered) by a pet male dog out in the yard until one day
poor, horny old Maud offered her backside to a strange
mongrel that proceeded to bite off her arm. More disturbing is Hamilton's description of a monkey named Jimmy
who one sunny afternoon discovered a human infant lying
in a hammock. "Jimmy promptly endeavoured to copulate
with the infant," observes Hamilton matter-of-factly. It's
unclear whether or not this was the author's own child,
nor is there any mention of the look on said human
infant's mother's face when she saw what Jimmy was
getting up to. In any event, though he may have had some
questionable child supervision skills, the candor with
which Hamilton reports on the sex lives of his monkeys
lends his tales that much more credence.

So why don't monkeys and apes masturbate nearly as
much as humans? It's a rarity even among low-status male
nonhuman primates that frustratingly lack sexual access to
females—in fact, the few observed incidents seem to be
with dominant males. And why haven't more researchers
noticed such an obvious difference with potentially
enormous significance for understanding the evolution of
human sexuality? After all, it's been nearly sixty years
since Alfred Kinsey first reported that 92 percent of
Americans were involved in masturbation leading to
orgasm.

The explanation for this cross-species difference, I'm convinced, lies in our uniquely evolved mental represent- ational abilities: we alone have the power to conjure up at will erotic, orgasm-inducing scenes in the personal movie theaters of our minds . . . internal, salacious fantasies com- pletely disconnected from our immediate external realities. One early sex researcher, Wilhelm Stekel, described masturbation fantasies as a kind of trance or altered state of consciousness, "a sort of intoxication or ecstasy, during which the current moment disappears and the forbidden fantasy alone reigns supreme."

Go on, put this aside, take a five-minute break, and put my challenge to the test (you may want to take leave for the restroom if you're on an airplane): try to masturbate successfully—that is, to orgasmic completion—without casting some erotic representational target in your mind's eye. Instead, clear your mind entirely, or think of, I don't know, an enormous blank canvas hanging in an art gallery. And of course no porn or helpful naked assistants are permitted for this task either.

How'd it go? If you're like most, you've seen the im- possibility of it. This is one of the reasons, incidentally, why I find it so hard to believe that self-proclaimed asexuals who admit to masturbating to orgasm are really and truly asexual. They must be picturing *something*, and whatever that something is gives away their sexuality.

Empirically capturing the phenomenology of mastur- bation fantasies is no easy matter. But some intrepid scholars have indeed tried to do so. In 1960, a British physician named Narcyz Lukianowicz, in an issue of the *Archives of General Psychiatry*, published one of the most sensational scientific reports I've ever had the pleasure of reading. Lukianowicz personally interviewed 188 people (126 males and 62 females) about their masturbation fantasies. An important caveat: All of these people were psychiatric patients with "various complaints and different neurotic manifestations," so their masturbation fantasies aren't necessarily typical. Nevertheless, the details provided by these patients about their erotic fantasies give us an extraordinary glimpse into the rich internal imagery accompanying human masturbation. Consider the self-report of a retired civil servant, aged seventy-one, being treated for obsessive feelings of guilt on account of his "excessive masturbation":

> I see in front of me naked beautiful women, dancing and performing some most exciting and tempting movements. After the dance they lean back, and keeping their legs wide apart, show their genitals and invite me to have sexual intercourse with them. They appear so real, that I can almost touch them. They're in a setting of an oriental harem, in a large oval room with divans and a lot of cushions around the walls. I can clearly see the wonderful gorgeous colors and the beautiful patterns of the tapestry, with an unusual vividness and with all the minute details.

Or consider Lukianowicz's account of a forty-four-year-old schoolmaster's fantasies, which reads like some bacchanalian, morphine-dappled scene ripped from the pages of William Burroughs's *Naked Lunch*:

> In them he "saw" naked adolescent boys with their penes stiffly erected, parading in front of him. As he progressed in his masturbation, the penes of the boys increased in size, till finally the whole field of his vision was filled with one huge, erect, pulsating penis, and then the patient would have a prolonged orgasm. This

type of homosexual masturbatory fantasy started shortly after his first homosexual experience, which he had had at the age of 10, and it persists unchanged hitherto.

Now, obviously, there are pathological cases of chronic masturbation where it actually interferes with the individual's functioning. In fact, it's not an uncommon problem for many caretakers of adolescents and adults with mental impairments whose charges often enjoy masturbating in public and making onlookers squeal and squirm in discomfort. (Not unlike some captive primates housed in miserable conditions such as laboratories or roadside zoos, where self-stimulation sometimes becomes obsessive.)

One thing that clinicians dealing with this problem may wish to consider is that the individual's cognitive limitations may not allow him to engage in more "appropriate" private masturbation because of difficulties with mental representation. In fact, frequency of erotic fantasies correlates positively with intelligence. The average IQ of Lukianowicz's sample was 132. So perhaps public masturbation, in which other people are physically present to induce arousal, is the only way that many with developmental disorders can achieve sexual satisfaction. Sadly, of course, society isn't very accommodating of this particular problem: between 1969 and 1989, for example, a single institution in the United States performed 656 castrations with the aim to stop the men from masturbating. One clinical study reported some success in eliminating this problem behavior by squirting lemon juice into the mouth of a young patient every time he pulled out his penis in public.

In any event, Lukianowicz argues that erotic fantasies

involve imaginary companions not altogether unlike children's make-believe friends. But unlike the more long-lived latter, he concedes, the former is conjured up for one very practical purpose: "As soon as the orgasm is achieved the role of the imaginary sexual partner is completed, and he is quite simply and quickly dismissed from his master's mind."

According to most findings in this area, men seem to entertain more visitors in their heads than do women. In a 1990 study published in *The Journal of Sex Research*, the evolutionary psychologists Bruce Ellis and Donald Symons found that 32 percent of men said that they'd had sexual encounters in their imagination with more than a thousand different people, compared with only 8 percent of women. Men also reported rotating in from their imaginary rosters one imagined partner for another during the course of a single fantasy more often than women did.

The psychologists Harold Leitenberg and Kris Henning summarized a number of interesting differences between the sexes in this area. In their review of research findings, the authors concluded that in general, a higher percentage of men reported fantasizing during masturbation than did women. It's important to point out, however, that neither "fantasy" nor "masturbation" was consistently defined across the studies summarized by Leitenberg and Henning, and some participants likely interpreted "masturbation" to mean simply self-stimulation (rather than orgasm inducing) or had a more elaborate conceptualization of "fantasy" than we've been using here, as some form of basic mental representation. For uncertain reasons, one dubious study compared "Blacks" and "Whites," so it's definitely a mixed bag in terms of empirical quality. They didn't find much of a difference, by the way.

A side note: both sexes claimed equally to have used their imaginations during intercourse. Basically, at some point, everyone tends to imagine someone—or something—else when they're having sex with their partner. There's nothing like the question "What are you thinking about?" to ruin the mood during passionate sex.

Here are some other interesting findings. Males report having sexual fantasies earlier in development (average age of onset 11.5 years) than do females (average age of onset 12.9 years). Females are more likely to say that their first sexual fantasies were triggered by a relationship, whereas males report having theirs triggered by a visual stimulus. For both men and women, straight or gay, the most common masturbation fantasies involve reliving an exciting sexual experience, imagining having sex with one's current partner, and imagining having sex with a new partner.

It gets more interesting, of course, once you step a little closer to the data. In one study with 141 married women, the most frequently reported fantasies included "being overpowered or forced to surrender" and "pretending I am doing something wicked or forbidden." Another study with 3,030 women revealed that "sex with a celebrity," "seducing a younger man or boy," and "sex with an older man" were some of the more common themes. Men's fantasies contain more visual and explicit anatomical detail (remember the giant, pulsating penis from Lukianowicz's study?), whereas women's involve more story line, emotions, affection, commitment, and romance. Gay men's sexual fantasies often include, among other things, "idyllic sexual encounters with unknown men," "observing group sexual activity," and—here's a shocker—images of penises and buttocks. According to one study, the top five lesbian fantasies are "forced sexual encounter," "idyllic encounter with established partner," "sexual encounters with men," "past gratifying sexual encounters," and—ouch!—"sadistic imagery directed toward genitals of both men and women."

One of the more intriguing things that Leitenberg and Henning conclude is that contrary to common (and Freudian) belief, sexual fantasies are not simply the result of unsatisfied wishes or erotic deprivation:

> Because people who are deprived of food tend to have more frequent daydreams about food, it might be expected that sexual deprivation would have the same effect on sexual thoughts. The little evidence that exists, however, suggests otherwise. Those with the most active sex lives seem to have the most sexual fantasies, and not vice versa. Several studies have shown that frequency of fantasy is positively correlated with masturbation frequency, intercourse frequency, number of lifetime sexual partners, and self-rated sex drive.

The authors also provide a fascinating discussion about the relation between sexual fantasy and criminality, including a clinical study in which deviant masturbatory fantasies were paired with the foul odor of valeric acid or rotting tissue. Now, that's enough to put a crimp in anybody's libido, I'd say. But Leitenberg and Henning's piece was written in 1995, summarizing even older research. This is important because it was still long before the mainstreaming of today's Internet pornography scene, where zero is left to the imagination.

And so I'm left wondering . . . in a world where sexual fantasy in the form of mental representation has become obsolete, where hallucinatory images of dancing genitalia, lusty lesbians, and sadomasochistic strangers have been replaced by a veritable online smorgasbord of real people doing things our grandparents couldn't have dreamed up even in their wettest of dreams, where randy teenagers no longer close their eyes and lose themselves to the oblivion and bliss but instead crack open their thousand-dollar laptops and conjure up a real live porn actress, what, in a general sense, are the consequences of liquidating our erotic mental representational skills for our species' sexuality? Is the next generation going to be so intellectually lazy in their sexual fantasies that their creativity in other domains is also affected? Will their marriages be more likely to end because they lack the representational experience and masturbatory fantasy training to picture their husbands and wives during intercourse as the person or thing they really desire? I'm not saying porn isn't progress, but over the long run it could turn out to be a real evolutionary game changer.

PART IV

Strange Bedfellows

Paedophiles, Hebephiles, and Ephebophiles, Oh My: Erotic Age Orientation

Michael Jackson, the late "King of Pop," probably wasn't a pedophile—at least not in the strict, biological sense of the word. It's a morally loaded term that has become synonymous with the very basest of evils. (In fact, it's hard to even say it aloud without cringing, isn't it?) But according to sex researchers, it's also a grossly misused term.

If Jackson did fall outside the norm in his "erotic age orientation"—and we may never know if he did—he was almost certainly what's called a hebephile, a newly proposed diagnostic classification in which mature adults display a sexual preference for children at the cusp of puberty, between the ages of roughly nine and fourteen. Pedophiles, in contrast, show a sexual preference for clearly *pre*pubescent children. There are also ephebophiles (from *ephebos*, meaning "one arrived at puberty" in Greek), who are mostly attracted to fifteen- to sixteen-year-olds; teleiophiles (from *teleios*, meaning "full grown" in Greek), who prefer those seventeen years of age or older; and even the very rare gerontophile (from *gerontos*, meaning "old man" in Greek), someone who has always been primarily aroused by the elderly (usually defined, at least for these purposes, as over sixty-five years of age). So although child sex offenders are often lumped into the single classification of pedophiles, biologically speaking it's a rather complicated affair. Some have even proposed an additional subcategory of pedophilia, "infantophilia," to distinguish those individuals most intensely attracted to children below six years of age.

Based on this classification scheme of erotic age orientations, even the world's best-known fictitious "pedophile," Humbert Humbert from Nabokov's masterpiece, *Lolita*, would more properly be considered a hebephile. (Likewise the protagonist from Thomas Mann's *Death in Venice*, a work that I've always viewed as something of the "gay *Lolita*.") Consider Humbert's telltale description of a "nymphet." After a brief introduction to those "pale pubescent girls with matted eyelashes," Humbert explains:

> Between the age limits of nine and fourteen there occur maidens who, to certain bewitched travelers, twice or many times older than they, reveal their true nature which is not human, but nymphic (that is, demoniac); and these chosen creatures I propose to designate as "nymphets."

Although Michael Jackson might have suffered disgrace from his hebephilic orientation, and his name will forever be entangled with the sinister phrase "little boys," he wasn't the first celebrity or famous figure who could be seen as falling into this hebephilic category. In fact, ironically, Michael Jackson's first wife, Lisa Marie Presley, is the product of a hebephilic attraction. After all, let's not forget that Priscilla caught Elvis's very grown-up eye when she was just fourteen, only a year or two older than the boys Michael Jackson was accused of sexually molesting. Then there's of course the scandalous Jerry Lee Lewis incident in which the twenty-two-year-old "Great Balls of Fire" singer married his thirteen-year-old first cousin.

In the psychiatric community, there's recently been much debate surrounding the issue of whether hebephilia, like pedophilia, should be designated as a medical disorder or, instead, seen simply as a normal variant of sexual orientation and not indicative of brain pathology. There are important policy implications of adding hebephilia to the checklist of mental illnesses, since doing so might allow people who sexually abuse pubescent children to invoke a mental illness defense. On the one hand, this defense would give perpetrators a medical excuse for their criminal behaviors. In most Western societies, most people are not entirely comfortable with this happening, because not only do they want the individual to be held accountable for his (or her) criminal actions, but a mental illness defense may also translate to the offender being treated at inpatient facilities rather than incarcerated in less-welcoming prisons. On the other hand, if hebephilia were regarded as a legitimate mental illness, such individuals could more easily be kept away from children indefinitely, since their civil liberties would be, in effect, absorbed by the state and they could therefore be kept institutionalized after serving their initial sentences. So a man who rapes a ten-year-old could more readily avoid prison because he's seen as having a certifiable, American Psychiatriac Association–backed mental disorder, but in the long run this is more likely to mean that he will never reenter society as a free citizen who "did his time."

One researcher arguing vociferously for the classification of hebephilia as a mental disorder is the psychologist Ray Blanchard. In an issue of *Archives of*

Sexual Behavior, Blanchard and his colleagues provide new evidence that many people diagnosed under the traditional label of pedophilia are in fact not as interested in prepubescent children as they are in early adolescents. To tease apart these erotic age orientation differences, Blanchard and his colleagues studied 881 men (straight and gay men recruited from the public) in his laboratory using phallometric testing (also known as penile plethysmography) while showing them visual images of differently aged nude models. Because this technique measures penile blood volume changes, it's seen as being a fairly objective index of sexual arousal to what's being shown on the screen—which, for those attracted to children and young adolescents, the participant might verbally deny being attracted to. In other words, the penis isn't a very good liar. So, for example, the image of a naked twelve-year-old girl (nothing prurient, but rather resembling a subject in a medical textbook) was accompanied by the following audiotaped narrative: "You are watching a late movie on TV with your neighbors' 12-year-old daughter. You have your arm around her shoulders, and your fingers brush against her chest. You realize that her breasts have begun to develop . . ."

Blanchard and his coauthors found that the men in their sample fell into somewhat discrete categories of erotic age orientation: some had the strongest penile response to the prepubescent children (the pedophiles), others to the pubescent children (the hebephiles), and the remainder to the adults shown onscreen (the teleiophiles). These categories weren't mutually exclusive. For example, some teleiophiles showed some arousal to pubescent children, some hebephiles showed some attraction to prepubescent children, and so on. But the authors did find that it's possible to distinguish empirically between a true pedophile and a hebephile using this technique, in terms of the age ranges for which men exhibited their strongest arousal.

They conclude that based on the findings from this study, hebephilia "is relatively common compared with other forms of erotic interest in children." Blanchard and his colleagues also argue that hebephilia should be added to the next version of the DSM (currently being revised) as a genuine paraphilic mental disorder—differentiating it from pedophilia. But not all of Blanchard's fellow scientists working in this area agree with this pathologizing approach. Most, in fact, are strongly opposed to conceptualizing hebephilia as a mental disorder. Their recalcitrance stems from the policy grounds mentioned earlier (we'll explore these in more detail below) but also from very basic, logistical concerns. The psychologist Thomas Zander points out that since chronological age doesn't always perfectly match physical age, including these subtle shades of erotic age preferences would be problematic from a diagnostic perspective: "Imagine how much more impractical it would be to require forensic evaluators to determine the existence of pedophilia based on the stage of adolescence of the victim. Such determinations could literally devolve into a splitting of pubic hairs."

There are also important theoretical reasons to question Blanchard's recommendation. Men who find themselves primarily attracted to young or "middle-aged" adolescents are social pariahs, since it's so strongly stigmatized, but historically (and evolutionarily) this wasn't necessarily the case. In fact, hebephiles—or at least ephebophiles—may have had a significant advantage over their competition. Psychologists have repeatedly found that markers of youth correlate highly, currently and historically, with perceptions of beauty and attractiveness. For straight men, this makes sense, since a woman's reproductive potential (and hence her "value" from a heartless evolutionary perspective) declines steadily after the age of about twenty. Obviously, having sex with a prepubescent child would be fruitless—literally. But, whether we like it or not, this isn't so for a teenage girl who has just come of age, who is reproductively viable, and whose brand-new state of fertility can more or less ensure paternity (therefore, being attracted to young girls represents a potentially powerful anticuckoldry strategy) for the male. These evolved motives have been portrayed unwittingly in many books and films, including the controversial movie *Pretty Baby*. In it, a young Brooke Shields played the role of twelve-year-old Violet, a prostitute's daughter in 1917 New Orleans whose coveted virginity goes up for auction to the highest bidder.

Understanding adult men's attraction to boys or adolescent males is more of an evolutionary puzzle; after all, it's not as though cuckoldry or reproductive years remaining is an issue here. But the psychologist Frank Muscarella's "alliance formation theory" attempts to unravel this homosexual age orientation. According to him, in the past, homoerotic affairs between older, high-status men and teenage boys served as a way for the latter to move up in ranks, a sort of power-for-sex bargaining chip. The most obvious example of this type of homo-sexual dynamic was found in ancient Greece, but some New Guinea tribes display these trends too. And of course, that desire which inspired Donatello's impish David still thrives, to say the least, in the world today. Just type the word *twink* (a slang term derived from that golden, cream-filled, phallic-shaped Hostess treat that describes a youthful gay male "with a slender, ectomorph build, little or no body hair, and no facial hair") in your Google image search bar and see what (or rather *who*) pops up. If you're bashful about doing that, there are plenty of more safe-for-work articles about these types of scandalous homosexual mentorships happening in Congress.

In any event, I'm guessing that Oscar Wilde would have signed on to Muscarella's theoretical perspective. After all, his famous "love that dare not speak its name" wasn't homosexuality per se, but rather a "great affection of an elder for a younger man,"

> as there was between David and Jonathan, such as Plato made the very basis of his philosophy, and such as you find in the sonnets of Michelangelo and Shakespeare. It is that deep, spiritual affection that is as pure as it is perfect. It dictates and pervades great works of art like those of Shakespeare and Michelangelo . . . It is beautiful, it is fine, it is the noblest form of affection.
>
> There is nothing unnatural about it. It is intellectual, and it repeatedly exists between an elder and a younger man, when the elder man has intellect, and the younger man has all the joy, hope and glamour of life before him. That it should be so, the world does not understand. The world mocks at it and sometimes puts one in the pillory for it.

But, in my opinion, Muscarella's theory doesn't pull a lot of weight. It addresses the erotic interests of the adult male in the relationship, sure, but it doesn't apply very well to the arousal patterns of teenage boys. Money, prestige, and status may make such affairs physically possible, and even symbiotic, as the author suggests. But as a general rule, gay teenage boys are aroused more by other teenage boys than they are by middle-aged men. Just as their male heterosexual counterparts grow up but still desire youthful female partners, gay boys simply turn into gay middle-aged men; their erotic preference for young partners doesn't change or go away either. And although there are exceptions, such as in ancient Greece, young males in most cultures have never seemed terribly interested in taking this particular route to success. Rather, and I may be wrong about this, since it's not the type of thing one experiments with these days, but I think most would prefer to scrub toilets for the rest of their lives or sell soft bagels at the mall than become the sexual plaything of an older gentleman.

In any event, given the biological (even adaptive) verities of adults being attracted to adolescents, most experts in this area find it completely illogical for Blanchard to recommend adding hebephilia to the revised diagnostic manual (especially since other, more clearly maladaptive paraphilias, such as gerontophilia, in which men are attracted to postmenopausal women, are not presently included). The push to pathologize hebephilia, argues the forensic psychologist Karen Franklin, appears to be motivated by "a booming cottage industry" in forensic psychology, not coincidentally linked with a "punitive era of moral panic." Because "civil incapacitation" (basically, the government's ability to strip a person of his or her civil rights in the interests of public safety) requires that the person be suffering from a diagnosable mental disorder, Franklin calls Blanchard's proposal "a textbook example of subjective values masquerading as science." Other critics similarly maintain that any such medical classifications based on erotic age orientations are rooted in arbitrary distinctions dictated by cultural standards.

One unexplored question, and one inseparable from the lightning-rod case that was Michael Jackson's molestation trials, is whether we tend to be more forgiving of a person's peccadilloes when we deem that individual as having some invaluable or culturally irreplaceable skills. For example, consider a true story, which I'll put first into the following general terms:

> There once was a man who fancied young boys. Being that laws were more lax in other nations, this man decided to travel to a foreign country, leaving his wife and young daughter behind, where he met up with another Westerner who shared in his predilections for pederasty, and there the two of them spent their happy vacation scouring the seedy underground of this country searching for pimps and renting out boys for sex.

If you're like most people, you're probably experiencing a shiver of disgust and a spark of rage. You may even feel these men should have their testicles drawn and quartered (halved?) by wild mares, be thrown to a burly group of rapists, be castrated with garden shears, or, if you're the pragmatic sort, be treated as any other sick animal in the herd would be treated, with a humane bullet to the temple or perhaps a swift and sure current of potassium chloride injected into a vein.

But notice the subtle change in your perceptions when I tell you that these events are from the autobiography of André Gide, who in 1947—long after he'd publicized these very details—won the Nobel Prize in Literature. Gide is in fact bowdlerizing his time in Algiers with none other than that great Dubliner wit, Oscar Wilde. Here is Gide's account:

> Wilde took a key out of his pocket and showed me into a tiny apartment of two rooms . . . The youths followed

him, each of them wrapped in a burnous that hid his face. Then the guide left us and Wilde sent me into the further room with little Mohammed and shut himself up in the other with the [other boy]. Every time since then that I have sought after pleasure, it is the memory of that night I have pursued.

It's not that we think it's perfectly fine for Gide and Wilde to have sex with minors or even that they shouldn't have been punished. (In fact, Wilde was sentenced in London to two years of hard labor for related offenses not long after this Maghreb excursion with Gide and died in penniless ignominy.) But somehow, as with many people's commingled feelings for Michael Jackson ("the greatest entertainer of all time") or perhaps even for the director Roman Polanski, the fact that these men were national treasures may dilute our moralistic anger.

For example, would you really have wanted Oscar Wilde euthanized like a lame animal because he fancied boys? Should André Gide, whom *The New York Times* hailed in its obituary as a man "judged the greatest French writer of this century by the literary cognoscenti," have been deprived of his pen, torn to pieces by illiterate thugs? (There's also Lewis Carroll's beloved *Alice's Adventures in Wonderland*, rumored to have been inspired by the author's hebephilic devotion to an eleven-year-old named Alice Liddell, not to mention the Italian painter Caravaggio's notoriously homo-erotic depictions of "fleshy, full-lipped, languorous young boys," as one critic put it.) It's complicated. And although in principle we know that all men are equal in the eyes of the law, just as we did for Michael Jackson during his bizarre legal affairs, I have a hunch that many other people also tend to feel (and uncomfortably so) a little sympathy for the Devil under such circumstances.

Whatever your feelings on this hot-button issue, one of the most significant challenges in studying people's erotic age orientation, from any theoretical perspective, is the fact that so many mainstream scientists are leery of commenting on this subject area or engaging with the research and debates surrounding the limited data available. Given that an overwhelming majority of child sex abuse cases involve male perpetrators, we would predict otherwise, but we still don't know, for example, whether measures of female genital arousal would show equivalent rates of pedophilia, hebephilia, and ephebophilia in women who were recruited from the general public.

My guess is that this academic unease is due in no small part to fear within the scientific community, since simply addressing the issue from an amoral perspective may be seen by some outraged segments of society as a type of pedophilia apologia. Frankly, I think such limbically fueled moral reactions are not only naive but shortsighted. You can't adequately address or change what you don't understand, after all. I've also a suspicion that all our fuming on this subject reveals something fairly significant about our sexuality. If there's one thing I've learned about human nature, it's that whenever society screams about some demon or another, it's probably just caught an especially alarming sight of itself in the mirror. And while not all men and women are attracted to adolescents, it's much more common a thing than we'd like to pretend. The bottom line is this: unless you're practicing mental gymnastics of the variety we explored in that earlier essay on masturbation, people haven't any say whatsoever about what their genitals respond to. *But people do have considerably more control over what exactly they do with these genitals. And at least in my book, those are very different things altogether.*

Animal Lovers: Zoophiles Make Scientists Rethink Human Sexuality

Out of context, some of our behaviors—if limited to the mere veneer of plain description—would raise many an eyebrow. The most innocent of things can sound tawdry and bizarre when certain facts and details are omitted. Here's a perfect example: I accidentally bit my dog Gulliver's tongue recently.

Now, you may be asking yourself what I was doing with his tongue in my mouth to begin with. But I would submit that that is a better question for Gulliver, since he's the one that violated my busily masticating maw by inserting that long, thin, delicatessen-slice muscle of his while I was simply enjoying a bite of a very banal bagel. Shocked by the feel of human teeth chomping down on his tongue, he yelped—then scampered off. Fortunately, Gulliver showed no signs of lasting trauma, and I was saved from having to explain to the vet how it came to be that I bit my dog's tongue off; but for days after the "incident" Gulliver kept his prized possession sealed behind the vault of his own clamped jaw. This gave my partner, Juan, and me at least a temporary reprieve from Gulliver's normally over-indulgent use of that particular organ on our faces. The story was strange enough for me to share with friends, and this particular tale of man bites dog unleashed the predictable onslaught of humorous bestiality innuendos. And that, ladies and gentlemen, is where the real story begins.

These sarcastic remarks from my confidants reminded me of a rather peculiar e-mail that I had received months earlier, written by an unusually erudite reader of my column. This individual was a self-professed "zoophile" (Greek for "animal lover") with a particular romantic affinity for horses, and he was hoping that I might write about this neglected, much-maligned topic of forbidden interspecies love. "The politics of acknowledging zoophilia as a 'legitimate' sexual orientation," asserted this reader, "often mean that zoophiles are either ignored as a class or subject to what can only be described as the most vicious, sustained, and hateful attacks by mainstream society."

I have my own viscerally based, unreasoned biases, and—I confess—on first reading this message, I promptly filed it away in the untouchable *Eww* . . . category of my mind. But Gulliver's tongue, combined with my sympathy for human underdogs, inspired me to go back and reread it, and I saw a rather intriguing scientific question lurking there. Is it really possible for an otherwise normal, healthy person to develop a genuine sexual *preference* for a non-human species?

Of course, there's nothing new under the sun about bestiality as a *behavior*. Prehistoric depictions of bestiality have been found in Siberia, Italy, France, Fezzan (in modern Libya), and Sweden. The ancient Greeks, Egyptians, Hebrews, and Romans allegedly partook in these sexual activities as well. Roman women were said to have inserted snakes into their vaginas and trained them to suckle from their nipples. Women allowed goats to enter them as part of some religious rituals in ancient Egypt.

Monkeys were once commonly trained to fondle men's genitals in the Nile and Indus valleys. But the act of having sex with an animal is one thing; being aroused *more* by animals than other humans is a different matter entirely. After all, the fact that I could, in principle, have sex with a woman—if I were plied with enough alcohol and she were tomboyish enough to create a suitable gender-modifying illusion—doesn't exactly make me a heterosexual. So it is with, say, a randy farm boy who finds himself one day with his phallus lodged curiously in a bucking goat, his eyes closed, and his brain replaying scenes of that flirtatious cheerleader from chemistry class. The act alone wouldn't make him a zoophile per se.

For decades, the scientific study of human beings' sexual relations with (other) animals has concentrated almost entirely on the overt act of bestiality, viewing such behavior as a surrogate for human-to-human sex. As a consequence of this approach, researchers have until very recently overlooked the possibility that some people might actually *favor* a romantic affair with a horse (or dog, lamb, cow, sow, or some other choice species) over the thought of becoming trapped in such unthinkable carnal relations with another person.

This emphasis on bestiality as a behavior rather than as a possible sexual orientation can be traced back at least as far as the work of Alfred Kinsey. In the classic book *Sexual Behavior in the Human Male*, Kinsey reported that 50 percent of the population of American "farm-bred males" claimed to have had "sexual contact"—he doesn't analyze further, so heaven only knows what behaviors these men who were raised on farms engaged in, exactly—with various other species, usually hoofed animals. Many of these people, said Kinsey, were ashamed of their early sexual experimentation with animals (most of these puerile encounters took place when the boys were between ten and twelve years of age), and so he advised clinicians to assure these now grown males that it was all part of being raised in a rural environment where females were scarce and premarital relations strictly forbidden. "To a considerable extent," wrote Kinsey, "contacts with animals are substitutes for heterosexual relations with human females."

But the stereotypical portrait of the zoophile as a woman-deprived, down-on-the-farm, and poorly educated male is being challenged by some contemporary findings. The most fascinating of these, in my opinion, is a set of two case studies published by the psychologists Christopher Earls and Martin Lalumière. The first case study documented the story of a low-IQ'd, antisocial, fifty-four-year-old convict who had a strong sexual interest in horses. In fact, this was why he was in prison for the fourth time on related offenses; in the latest incident, he had cruelly killed a mare out of jealousy because he thought she'd been giving eyes to a certain stallion. (You thought *you* had issues.) The man's self-reported sexual interest in mares was actually verified by a controlled, phallometric study. When he consented to be hooked up to a penile plethysmograph in prison and was shown nude photographs of all varieties and ages of humans, the man was decidedly flaccid. Nothing happening down there either when he looked at slides of cats, dogs, sheep, chickens, or cows. But he certainly wasn't impotent, as the researchers clearly observed when the subject was shown images of horses.

This case and related anecdotal evidence reported by the authors (including a 1950s study of a sixteen-year-old "imbecile" who sexually preferred rabbits to women) were important at the time because they suggested that zoophilia may be an extraordinarily rare—but real—type of minority sexual orientation. That is to say, for some people, having sex with their animal "lovers" may amount to more than just replacing human sex with the next-best thing. Rather, for them, sex with nonhuman animals is the best thing.

On the heels of their study in 2002, Earls and Lalumière report having received a number of letters and e-mails from people who also self-identified as zoophiles (or "zoos," as many of these individuals refer to themselves on the Internet, which has served to connect them in un- precedented ways and to attract curious researchers like flies on a barnyard wall). And many of these respondents were vehement that they didn't fit the mentally challenged, rural stereotype reflected by Kinsey's analysis. Some were, in fact, highly educated professionals. But what most con- cerned these people was society's misconception that they were somehow harming animals by being amorous with them. The majority of zoophiles scoffed at the notion that they were abusive toward animals in any way—far from it, they said. Many even considered themselves to be animal welfare advocates in addition to being zoophiles.

In an effort to disentangle myth from reality, then, Earls and Lalumière published a new case study focusing on the first-person account of a forty-seven-year-old, high- functioning (he earned his M.D. at age twenty-eight), and seemingly well-adjusted male who had had, by all appear- ances, a completely unremarkable city upbringing with loving parents and no memories of abuse or neglect. Nonetheless, from an early age this man had struggled to come to grips with his own zoophiliac tendencies. Again, horses served as the primary erotic target:

As I grew into adolescence my sexual ideation was different from what it was supposed to be. I looked at horses the same as other boys looked at girls. I watched cowboy movies to catch glimpses of horses. I furtively looked at pictures of horses in the library. This was before the Internet and I felt totally isolated. I was a city boy. I had never seen a horse up close, never touched or smelled one. No one in my family had any contact with horses, but for me, they held a powerful, wonderful, and, yes even—well, primarily—sexual attraction. I had no idea that there were others like me in the world. I tried to be normal. I tried to get interested in girls, but for me they were always foreign, distasteful and repulsive. A couple of early adolescent sexual explorations ... were mechanical, forced and unsuccessful.

At the age of fourteen, the boy had managed to find the nearest horse stables, which he would visit frequently—secretly—by bicycle. Imagine him there, a young boy lurking in the fields, leaning against fencing in the meadows, perhaps under the strawberry–pale blue sky of early autumn, longing to be close to these huge, mysterious creatures that created such strange stirrings in his loins. Eventually, they came close enough for him to touch them and smell them, a scent he would describe over thirty years later as "astonishingly wonderful." This was no copycat version of the fabled play *Equus* (in fact, it was still years before the alleged British case of bestiality that the play was loosely based on) but instead a real developmental experience for an otherwise normal human being. Three years later, the teenager purchased his own mare, took riding lessons, and began a "long courtship" with the female horse until, finally, the pair consummated their relationship:

When that black mare finally just stood there quietly
while I cuddled and caressed her, when she lifted her tail
up and to the side when I stroked the root of it, and
when she left it there, and stood quietly while I climbed
upon a bucket, then, breathlessly, electrically, warmly, I
slipped inside her, it was a moment of sheer peace and
harmony, it felt so right, it was an epiphany.

This case study reveals that, again, it's not only mentally
deficient farmhands who have sex with animals. And
neither, it seems, is it simply unattractive, unsavory men
who can't find willing sexual partners of their own species.
In fact, shortly after obtaining his medical degree, this
particular man married a (human) woman and had two
children with her. But their sex life relied on his imagining
her to be a horse, and—perhaps not surprisingly—the
marriage didn't last. As my sister said when I mentioned
this tidbit to her: "I can see how that would be a
problem."

Another pioneering researcher in zoophilia, the
Maryland-based sexologist Hani Miletski, found similarly
that more than half of the ninety-three self-identified
zoophiles she'd spoken to (eighty-two men and eleven
women with an average age of thirty-eight) reported being
more attracted to animals than to people. And just like the
mare lover from Earls and Lalumière's study, the majority
(71 percent) considered themselves to be well adjusted in
their current lives, with 92 percent seeing no reason to stop
having sex with their animal partners. This is an important
point, because the current version of the American
Psychological Association's DSM-IV classifies zoophilia as
a disorder only if an individual's sexual attraction to nonhuman animals causes the person to experience distress.
Bestiality is still illegal in most states, but it's rarely

prosecuted, mainly because it's quite a challenge catching an interspecies coital coupling as it's happening.

As you can probably imagine, though, the subject of zoophilia is a highly charged one, attracting the ire of animal rights groups such as People for the Ethical Treatment of Animals and causing a knee-jerk moralistic response in the rest of us platonic animal lovers. Ironically, it landed one prominent animal rights defender, the Princeton philosopher and author Peter Singer, in some hot water. In an essay for *Nerve* magazine called "Heavy Petting," Singer was asked to review *Dearest Pet* by the Dutch biologist Midas Dekkers. But he did more than just review the book. Singer also asked readers to reconsider whether humans having mutually pleasurable, nonabusive sex with other animals is as inherently wrong as we've been led to believe by our traditional Judeo-Christian mores (go on, quote Leviticus). As Singer noted, "The vehemence with which this prohibition [against sex with other species] continues to be held, its persistence while other non-reproductive sexual acts have become acceptable, suggests that there is [a] powerful force at work: our desire to differentiate ourselves, erotically and in every other way, from animals."

Singer told me that he wasn't advocating sex with animals, but rather just raising the question of why we find it so objectionable. Ever since, however, the piece has been used against Singer by his opponents, most of whom are trying to discredit his controversial views on human euthanasia and abortion. "Look," argue many of Singer's critics, "how can we take anything this guy says seriously when he wants us to have sex with animals!" But most zoophiles, of course, tend to agree with Singer's general assessment of human "speciesism" being cloaked under the tenuous justification of animal protection. After all, we *are* animals.

In a chapter published in *Transgressive Sex: Subversion and Control in Erotic Encounters*, the anthropologist Rebecca Cassidy offers a particularly sad account of how this religiously laden assumption that humans are "more than animal" manifested itself in a 1601 courtroom in Rognon, France. It was there that a sixteen-year-old girl named Claudine de Culam was being tried for bestiality with her pet dog:

> Apparently uncertain as to whether such an act was anatomically possible, the judge appointed a number of female assistants in order to put the dog and the girl to the test. As the women undressed Claudine, the dog leaped upon her. On the basis of this evidence both the dog and the young woman were strangled, their bodies burned and scattered to the four winds, "that as little trace as possible might remain to remind mankind of their monstrous deeds."

One especially provocative study published in the *Archives of Sexual Behavior* involved the sociologists Colin Williams and Martin Weinberg attending a zoophile gathering on a farm, where a group of predominantly young men—nearly all of whom were college educated—were observed to have "genuine affection" for the animals they had sex with. Many zoophiles consider "zoosadists" anathema, and they have been sincerely striving to distance themselves from those who derive pleasure from harming animals. Yet some scholars, such as the criminologist Piers Beirne, contend that zoophiles incorrectly assume that animals are capable of consenting to having sex with them, and therefore human sexual relations with animals of any kind should be considered "interspecies sexual assault."

In taking stock of my own position on this touchy subject, I find myself emotionally drawn to Beirne's "zero tolerance" stance. If some unscrupulous zoophile were to lure away my beloved dog Uma with a bacon strip into the back of his van, well, hell hath no fury—even if she did come back wagging her tail. But this is mostly just the reflexive moralizer in me. Words such as *pervert* and *unnatural* have all the theoretical depth of a thimble. Rationally, Singer is right to question our visceral aversion to interspecies sex. And having had a chimpanzee in estrus forcibly back her swollen anogenital region into my midsection ("Darling," I said, "not only are you the wrong species, but the wrong sex") and more dogs than I care to mention mount my leg, I know that it's not only humans who are at risk of misreading sexual interest in other species. The Arabian stallion that impaled a Seattle man with its erect penis in 2005, fatally perforating the man's colon, makes one wonder who the victim really was.

And if zoophilia occurs in certain members of our own species, could members of other species be aroused primarily by humans? In Maurice Temerlin's book *Lucy: Growing Up Human*, the author claims that once his chimpanzee "daughter" reached sexual maturity, the chimp was interested sexually only in human males. Temerlin, a psychotherapist, even bought Lucy a *Playgirl* magazine and found her rubbing her genitals on the full-page spread of a naked man.

In any event, philosophical questions aside, I simply find it astounding—and incredibly fascinating from an evolutionary perspective—that so many people (as much as a full 1 percent of the general population) are certifiable zoophiles. And scientific researchers appear to be slowly conceding that zoophilia may be a genuine human sexual orientation. Still, just as you probably do, I have a slew of unanswered questions that have yet to be addressed empirically. What makes some domestic species—such as horses and dogs—more common erotic targets for zoophiles than others, such as, say, cats, llamas, or pigs? (Clawed cats would be a problem.) Do zoophiles find particular members of their preferred species more "attractive" than other individuals from those species, and if so, are they seduced by standard beauty cues, such as facial symmetry in horses? What is the percentage of homosexual zoophiles (those who prefer animal partners of the same sex) over heterosexual zoophiles? Aside from the hoof marks on their foreheads, how do zoophiles differentiate between a "consenting" animal partner and one who isn't "in the mood"? Why are men more likely to be zoophiles than women? Are zoophiles attracted only to sexually mature animals—and if not, does this make them "zoopedophiles"? What about cross-cultural differences? Is the tendency to become a zoophile heritable?

We'll have to wait a while longer for some intrepid sexologist to dig into these and other unanswered scientific questions about zoophilia, perhaps the rarest of all the sexual paraphilias. Meanwhile, I must confess that I'm a bit jealous of you caring zoophiles out there. How nice it would be to be able to dispense with all those emotional encumbrances that come with being attracted to other members of the human species. If only I could settle down discreetly with a sassy little bitch—a consenting adult, of course—life would probably be a lot easier.

Asexuals Among Us

Gay people are often asked by the curious straight: "When did you first realize you were gay?" In my case, I remember undressing my Superman doll—and being terribly disappointed at the result—as well as being motivated to befriend the more attractive boys in third grade. But hormonally speaking, it wasn't until I was about fourteen that I first looked in the mirror and thought, Ah, that's what I am all right, it all makes perfect sense now.

It wasn't that much of a mystery. After all, lust isn't exactly a subtle thing. Back then, I derived as much pleasure from making out with my "girlfriend" as I might have from scraping the plaque from my dog's teeth. In contrast, barely touching legs with a boy I had a crush on sparked an electric, ineffable ecstasy. In the locker room after high school gym class, I forced myself to picture naked girls in my head (particularly my girlfriend) as a sort of cognitive cold shower, a preemptive strike against an otherwise embarrassing physical response. I could go on, but you get the idea: whether or not we like, hide, or accept what we are, our true identities—gay, straight, bisexual—consciously dawn on each of us at some point in our lives, usually by adolescence. We all have a natural "orientation" toward sexual contact with others, and for the most part we're just hopeless pawns, ineffectual onlookers, to our bodies' desires.

At least, that's what most people tend to think. But some scientists believe that there may be another sexual orientation in our species, one characterized by the absence of desire and no sexual interest in males or females, only a complete and lifelong lacuna of sexual attraction toward any human being (or nonhuman being). Such people are regarded as asexuals. Unlike bisexuals, who are attracted to both males and females, asexuals are equally indifferent to and uninterested in having sex with either gender. So imagine being a teenager waiting for your sexual identity to express itself, waiting patiently for some intoxicating spurt of lasciviousness to render you as dumbly carnal as your peers, and it just doesn't happen. These individuals aren't simply celibate, which is a lifestyle choice. Rather, sex to them is just so . . . boring.

In one study from 2007, a group of self-described adult asexuals was asked how they came to be aware they were different. One woman responded:

> I would say I've never had a dream or a fantasy, a sexual fantasy, for example, about being with another woman. So I can pretty much say that I have no lesbian sort of tendencies whatsoever. You would think that by my age I would have some fantasy or dream or something, wouldn't you? . . . But I've never had a dream or a sexual fantasy about having sex with a man, either. That I can ever, ever remember.

In another study, an eighteen-year-old woman put it this way:

> I just don't feel sexual attraction to people. I love the

human form and can regard individuals as works of art and find people aesthetically pleasing, but I don't ever want to come into sexual contact with even the most beautiful of people.

According to the psychologist Anthony Bogaert, there may be more genuine asexuals out there than we realize. In 2004, Bogaert analyzed survey data from more than 18,000 British residents and found that the number of people (185, or about 1 percent) in this population who described themselves as "never having a sexual attraction to anyone" was just slightly lower than those who identified as being attracted to the same sex (3 percent). Since this discovery, a handful of academic researchers have been trying to determine whether asexuality is a true biological phenomenon or, alternatively, a slippery social label that for various reasons some people may prefer to adopt and embrace.

Sexual desire may wax and wane over the course of a life or—as many people on antidepressants have experienced—become virtually nonexistent because of medications or disease. There are also chromosomal abnormalities, such as Turner's syndrome, often associated with an absence of sexual desire. Traumatic events in childhood, such as sexual abuse, can factor into an aversion to sex too. But if it exists as a distinct orientation, true asexuality would be due neither to genetic anomaly nor to environmental assault. And indeed, although little is known about its etiology (Bogaert believes it may be traced to prenatal alterations of the hypothalamus), most asexuals are normal, healthy, hormonally balanced, and sexually mature adults. For still uncertain reasons, they've just simply always found sex to be one big, bland yawn. Asexuality would therefore be like other sexual orientations in the sense that it is not "acquired" or "situational" but rather an essential part of one's biological makeup. Just as a straight man or a lesbian can't wake up one day and decide to become attracted to men, neither could a person—in principle, anyway—"become" asexual. Sexual dysfunctions such as hypoactive sexual desire disorder can also be ruled out if a "preference" toward a gender does not awaken in response to clinical intervention such as hormonal treatment. As Bogaert notes, even those with object fetishes or paraphilias usually display a gender-based attraction, such as men who have a thing for women's shoes or necrophiliacs who have sex with dead women's (but not men's) bodies.

But the story of asexuality is very complicated. For example, as discussion on the AVEN (Asexual Visibility and Education Network) website forums demonstrate, there is tremendous variation in the sexual inclinations of those who consider themselves asexual. Some masturbate; some don't. Some are interested in nonsexual, romantic relationships (including cuddling and kissing but no genital contact), while others aren't. Some consider themselves to be "hetero-asexual" (having a nonsexual aesthetic or romantic preference for those of the opposite sex), while others see themselves as "homo-" or "bi-asexuals." There's even a matchmaking website for sexless love called Asexual Pals. Yet many asexuals are also perfectly willing to have sex if it satisfies their sexual partners; it's not awkward or painful for them, but rather, like making toast or emptying the trash, they just don't personally derive any pleasure from the act. As the researchers Nicole Prause and Cynthia Graham found in their interviews with self-identified asexuals, "They were not particularly sexually fearful . . . they had a lower excitatory drive." Others insist on being in completely sexless relationships, ideally with other asexuals. Thus, while many asexuals are virgins, others are ironically even more experienced than your traditionally sexual friends. Some want children through artificial means such as in vitro fertilization; others are willing to have them the old-fashioned way or don't want children at all.

Thus, on the one hand, there seems to be a sociological issue of people of a marginalized sexual identity gathering steam and beginning to form an identifiable community (and in the process attracting significant media attention, including coverage on *The Montel Williams Show* and *The View* and a feature story in *New Scientist*). On the other hand, there remains—to me—the more intriguing biological issue of asexual essentialism; that is to say, is it really possible to develop "normally" without ever experiencing sexual desire, even a niggling little blip on the arousability radar, toward any other human being on the face of the earth? I have little doubt that there are self-identified asexuals who would fail to meet this essentialist criterion, but if even a sliver of the asexual community has truly never experienced arousal, then this would pose fascinating questions for our understanding of human sexuality and evolutionary processes.

Scientists have just scratched the surface in studying human asexuality. You can count the number of studies on the subject on one hand. So questions remain. Does asexuality, like homosexuality, have heritable components? Certainly that's plausible. After all, historically, at least female asexuals, who wouldn't need to orgasm to conceive, would have probably had offspring with their male sexual partners, thus ensuring continuity of the genetic bases of asexuality. (Although Bogaert's original findings suggested that asexuality was somewhat more common among women, more recent research by Prause and Graham found no such gender difference in their college-aged sample of self-reported asexuals.) If some asexuals masturbate in the absence of sexual fantasy or porn, then what exactly is it that's getting them physically aroused? (And how does one achieve orgasm—as some asexuals apparently do—without experiencing pleasure?) Also, if you're on board theoretically with evolutionary psychology, almost all of human cognition and social behavior somehow boils down to sexual competition. So what would the evolutionary psychologist make of asexuality? If sex is nature's feel-good ruse to get our genes out there, is there actually a natural category of people that is immune to evolution's greatest gag?

I have to say the only good way to solve the riddle is also a bit unsavory. But unless psychological scientists ever gather a group of willing, self-identified asexuals and, systematically and under controlled conditions, expose them to an array of erotic stimuli while measuring their physical arousal (penile erection or vaginal lubrication), the truth of the matter will lie forever hidden away in the asexuals' pants.

Foot Play: Podophilia for Prudes

We've discussed pedophilia already, but let's talk about podophilia, the love of feet and often, by extension, shoes. Actually, there's a fair share of podophiliac pedophiles, so it's worth pointing out that the two are not mutually exclusive. But in any event, at the risk of veering off already into a fetish of an entirely different sort (acroto-mophilia, which is the love of amputees or, more specifically, a lusting after their severed limbs), let me begin by saying that I've always found feet—those elongated, malodorous, beknuckled terrestrial hands—somehow awfully off-putting. Not that I'd prefer them snipped off my partners, but you know what I mean.

In fact, my own distaste for feet makes podophilia all the more intriguing to me because, among other things, it goes to show how receptive to learning our carnal taste buds may in fact be in contributing to what later becomes delectable. Perhaps my genitals only lacked a mysterious encounter with other people's feet during a critical period of my sexual development. Many people who derive their primary sexual satisfaction from "foot play" can trace their fondness for feet to specific episodes either in their childhoods or during their early adolescence.

One of the most important and detailed historical treatments of the subject of foot (and shoe) fetishism is that by the British sexologist Havelock Ellis in 1927. "In a small but not inconsiderable minority of persons," writes Ellis, "the foot or boot becomes the most attractive part of a woman, and in some morbid cases the woman herself is regarded as a comparatively unimportant appendage." Ellis describes the case of the eighteenth-century French novelist Rétif de la Bretonne, whose irreverent literary works were filled with references to his own fancies. (In fact, the eponymous *retifism* is an arcane term for foot fetishism.) In Rétif's very frank autobiography, *Monsieur Nicolas*, the then-sixty-year-old writer reminisces about being smitten with a girl's feet as early as age four. Rétif's theory about the origins of his foot fetish was that since feminine freshness and purity were so prized in his day, those ladies who managed to keep that part of their body which met directly with the dirt so delicate and unspoiled were the most attractive of all.

"This taste for the beauty of the feet," reflects Rétif of his upbringing in the Burgundy region of France, "was so powerful in me that it unfailingly aroused desire . . . When I entered a house and saw the boots arranged in a row, as is the custom, I would tremble with pleasure; I blushed and lowered my eyes as if in the presence of the girls themselves." What was especially arousing to Rétif, Ellis explains, was his knowledge that these objects had absorbed the essence of the feet he so desired. "He would kiss with rage and transport whatever had come in close contact with the woman he adored." In fact, he wished desperately to be buried with the "green slippers with rose heels and borders" of an older woman whose feet he'd become infatuated with as a teenage boy.

More recent work has validated Ellis's hunch that shoe fetishism is not simply a peculiar attraction to these inanimate objects and that sexual arousal is related to the intimate connection with the feet of a particular shoe owner. For example, in a series of reports on male homo-sexual foot fetishism, the sociologist Martin Weinberg and his colleagues asked members of the Foot Fraternity what they found especially attractive in shoes. The majority of these 262 men expressed complete sexual disinterest in new, never-worn shoes. Instead, they had a clear prefer-ence for footwear that had been well worn by a good-looking person. Shopping for shoes at thrift stores was a godsend for many of these fetishists, since it allowed them to fantasize about the original owner rather than face the ugly realties of an aesthetically challenged sole—*soul*, I mean *soul*. And just as heterosexual podophilia has a symbolic element, with straight connoisseurs displaying very particular tastes for certain styles of female footwear and leggings, gay foot fetishists associate shoe types with idealized males. One man, for example, explained to the investigators how a rich tapestry of senses had become linked to stereotypical associations with different male shoe types: "the odors and the corresponding image; docksiders and preppies, sneakers and young punks, boots and dominant men." Other gay male foot fetishists repeated this symbolic theme in their likes and dislikes:

"Boots represent power and strength ... They exemplify the essence of manhood, an exaggeration of maleness."

"Wing tips typify a successful businessman."

"Sneakers have been in contact with a good-looking young stud."

"Weejun penny loafers are worn by preppie college guys."

In a subsequent article in *The Journal of Sex Research*, Weinberg and his colleagues returned to their foot fan base (the Foot Fraternity boasted more than a thousand members in 1995, the vast majority being well-educated white men with white-collar jobs) and asked these individuals to reflect in writing on the origins of their love of male feet. "We specifically asked the age at which respondents first became interested in feet/footwear," explain the authors, "and, to tap the reinforcing effect of masturbation"—no pun intended, I think—"their experiences with fantasies about feet/footwear when they masturbated during adolescence." Fetishists reported a mean age of twelve in their first becoming (consciously) sexually aroused by feet, with nearly all of them masturbating regularly to foot-related objects (such as shoes or socks or photographs of feet) or to mental images of steamy podiatric encounters.

In terms of developmental context, many of the 204 respondents couldn't recall a specific incident per se from their pasts that they might attribute to this now-concretized aspect of their adult sexual identity. Yet 89 were able to give detailed accounts of their suspected first foot-related triggers. And for you parents out there with your toes dangling promiscuously for your impressionable young children to see, their answers might give you pause. "Sleeping upside down with my parents," reflects one grown man of his early childhood and his snuggling innocently with them under the covers, "and finding my dad's feet in my face." "I used to tickle my dad's feet," recalls another. "I enjoyed his laughter very much . . . He would feign enjoyment as part of the game." Another reminisces: "At about 5 or 6 years old, removing my father's shoes and massaging his hot feet ... The soft, warm feet and the pleasure he seemed to experience—usually going to sleep—and I could kiss and lick his feet." Other respondents had similar experiences, but theirs were unrelated to parental feet. The foot of a hero-worshipped older brother hanging down in front of one's face as he lay on a topmost bunk bed, for example, or wrestling playfully with one's friends or neighbors and finding a foot buried, and not unpleasantly so, in one's crotch.

As with Ellis's analysis of heterosexual podophiles, Weinberg and his colleagues observed that the origins of this homosexual podophilia could almost always be traced back to such positive experiences during development, rather than negative or abusive ones. This is an important observation, in fact, because it's often assumed that such a fetish represents the person's masochistic desire to be kicked or violently tread upon. Although this is in some cases true, Ellis admonishes us not to so hastily assume that the average foot lover has a secret desire to be subservient to a dominant figure. "To suppose that a fetishistic admiration of his mistress's foot is due to a lover's latent desire to be kicked," he proclaims, "is as unreasonable as it would be to suppose that a fetishistic admiration for her hand indicated a latent desire to have his ears boxed."

Ellis was convinced that it's often the most intelligent and precocious children who are particularly "liable to become the prey of a chance symbolism" in their sexual development, forever shaping their adult orientations. One especially vivid example of such a child, in this case a very troubled one, is laid out in the *American Journal of Psychotherapy* in an article titled "The Treatment of a Child Foot Fetishist." A team of physicians led by Jules

Bemporad handled the case. The boy, whom the psychiatrists called "Kurt," presented initially to the children's clinic at age eight. His full-scale IQ tested in the superior range at 129, but somewhere along the way he'd acquired the bizarre habit of sneaking up on his mother, removing her shoes, and licking her feet with great excitement. "While licking the feet," write the psychiatrists, "he regularly had an erection and played with his penis." Digging a bit deeper into the boy's past, the following tale emerged:

> The preoccupation with his mother's feet began within the first year of life; the mother remembers that he "loved to play with my feet" and that she encouraged it, considering it cute. She would lie in bed while Kurt gave her a foot massage—an enjoyable experience for her and a source of comfort for him. Gradually, the rubbing was accompanied by mouthing and licking, as well as by the mother giving him monetary rewards for his "massages." By the age of 5 or 6, the act had become sexually exciting, leading to wild shouting and genital manipulation. It was at this stage that the mother allegedly began prohibiting contact with her feet.

At this stage, of course, it was too late. The authors follow Kurt up to age sixteen. While he continued to excel at school and managed to gain control over his blatant symptoms with his mom's toes, his foot obsession remained very much intact, and his mother's playful per- missiveness left lifelong sexual problems. There were other factors involved, too, that made for a Freudian nightmare. Kurt's distant and anal-retentive Jewish father allegedly informed the boy, on passing a delicatessen one day, that the salamis hanging in the storefront window were the severed penises of dead men. (To retaliate, Kurt started decorating his room with trappings of Christianity.) And the mom admitted to playing with her little son's penis during baths and calling it "cute." About a decade later, the British child psychotherapist Juliet Hopkins would describe the case of a very tomboyish little girl who also had a problematic eroticized interest in feet. Hopkins's interpretation of the origins of the girl's foot fetish is that it all started in the bathtub. Her father used to bathe the little girl while she sat in his lap in the tub. From the child's perspective, says Hopkins, seeing the two sets of feet together with their similar appearance was comforting and empowering to the girl because it offset the more obvious —and threatening—difference in genital anatomy.

Still, while sensational stories are easy to come by, it's only the slim minority for whom this erotic penchant for feet turns sinister or criminal. Most psychiatrists believe that unless it interferes with the individual's adjustment in society or his or her mental well-being, fetishes shouldn't be treated as a "problem" requiring clinical intervention. Eighty percent of Weinberg's homosexual sample, in fact, reported being in a relationship with an understanding partner who was willing to accommodate their unshared fetish by incorporating foot play into the couple's normal sexual routines. (In fact, given his noticeable interest in fellating my toes, looking back, I suspect that one of the first men I was ever with had a secret foot fetish. Honestly, I wouldn't have much minded; I shooed him away only for his own good, since I *did* have a nasty case of athlete's foot that summer.) Related to partner support, the researchers also found that having access to member groups such as the Foot Fraternity significantly reduces confusion and discontent, allowing like-minded individuals to come out of the closet—or out of the shoe box—and explore their shared interests in open comfort within a nonstigmatizing community.

This live-and-let-live approach certainly wasn't the tack of the therapist Joseph Cautela in 1986, however. Cautela submitted an actual case transcript to the *Journal of Behavior Therapy and Experimental Psychiatry* detailing his first session with a very lonely thirty-one-year-old foot fetishist who, ever since roughhousing with other boys when he was a teenager and becoming aroused by all the flying feet, found himself fantasizing about the feet of twelve- to fourteen-year-old boys. Importantly, the man had never acted on these feelings, but he wanted to be "normal" and so sought treatment. Cautela attempted to reorient the patient, trying to turn him not only off boys' feet but from the male sex entirely. Of course few parents would be thrilled to discover this fellow employed at Kids Foot Locker, but from the case report he at least appeared to be pretty harmless, so his treatment is a rather sad testament to those times. But you be the judge. Let's listen in to their therapy session:

THERAPIST: It's very important for you to know that every time you fantasize and masturbate about wrestling with boys you make the fetish worse. It's just like doing it in reality. You strengthen the habit.
PATIENT: I guess you're right, but it's out of control.
THERAPIST: Well, I'll help you get control of the habit.
PATIENT: Can you?
THERAPIST: Well, we have a good chance if you cooperate. I can teach you relaxation, teach you the self-control triad to get rid of your negative thinking and have you imagine something terrible or disgusting happening if you start inappropriate sexual fantasy.

PATIENT: Is that all?

THERAPIST: No. There are other coping mechanisms we can use. Also, we have to try to get you aroused toward females.

PATIENT: But isn't that a sin?

THERAPIST: Well, what is more sinful: having a foot fetish that can ruin your life, or learning to be aroused by females?

PATIENT: Well, if you put it that way.

THERAPIST: I'm just saying that, in my experience in treating fetishes, it is necessary to build up heterosexual relationships and arousal. It's up to you if you want to change. That's my approach.

PATIENT: OK. That makes sense.

Heterosexual podophiles are difficult enough to explain from an evolutionary perspective. Under certain conditions in the ancestral past, such male foot fetishists (among perhaps other fetishists) may have, strangely enough, had a leg up over those whose arousal patterns were less discriminating. Most fetishists are known to have very specific tastes, and so partners matching their desires and willing to accommodate them—or, in this case, possessing feet that make them go crimson—are hard to find. Yet, in some cases, having fewer reproductive partners and instead having sex with only very particular females may be the key to success.

This is the intriguing, if speculative, theory hinted at by the researcher James Giannini and his colleagues in *Psychological Reports*. It seems that, historically, the cultural eroticization of the female foot has coincided with the presence of sexually transmitted epidemics in such cultures. Podophiliac tastes have waxed and waned as diseases have run their course, and the authors illustrate how foot love manifested itself, then subsided, during the gonorrhea epidemic in the thirteenth century, syphilis in the sixteenth and nineteenth centuries, and AIDS in the current century. In sixteenth-century Spain, for instance, painters began specializing, for the first time in history, in portraits of the female foot, and shoes that showed a teasing bit of "toe cleavage" were all the rage. Again, Giannini's ideas here are highly speculative, but it's a promising hypothesis waiting to be borne out by additional population-level data on sexual behaviors and fetishism. If the shoe fits, as they say.

A Rubber Lover's Tale

On June 6, 1969, a detective in southern Michigan, apparently sensing some scholarly significance in the unusual case report before him, sat down at his desk and typed up a matter-of-fact, single-page cover letter to an associate at the Kinsey Institute for Sex Research. The detective was writing with regard to a male patient who was being held voluntarily at a Kalamazoo psychiatric ward—a polite, self-confessed "rubberphile" who, in the darkest burrows of his own deep shame and mortification, with the electric summer hum of cicadas, the shrill of rusted gurney wheels, and the groans of fellow patients as an orchestra for his thoughts, had for several long weeks before sat hunched over in his bed trying furiously to expurgate his sexual demons through his pen. "This report is my soul and will save my life," wrote the patient. And it's this report that came to land soon after on the detective's desk and was looked at askance, stuffed in a manila envelope, borne off by airmail to Bloomington, and eventually shelved discreetly with tens of thousands of other such reports in the Kinsey Institute's unpublished archives.

Forty years later, under the soft glow of fluorescent lighting in the institute's library, I happened across this fetishist's handwritten sexual autobiography—along with the detective's austere covering note—while working on a book, and I must say that this man's presentation of his condition was an articulate, startling self-exorcism. In a document still effervescent with fear and spanning some fifty pages of lucid, densely packed prose glazed with biblical scripture, this tortured forty-one-year-old "rubber lover"—who'd been arrested for various rubber-related crimes, the most minor of these being his making thousands of indecent phone calls to department store saleswomen, inquiring about rubber bikinis for his imaginary wife while fondling laminated advertisements of elastic-clad models with one hand and himself with the other—worked feverishly to understand the origins of his own insatiable desire for rubber and flesh.

To the best of his knowledge, it all started when, at the age of seven, he'd stumbled upon his mother's glistening white rubber bathing suit hanging on a clothesline on the back porch, an arousing event that coincided with his first becoming aware of that strange stirring in his loins. What began as an innocent enough youthful quirk, however, would eventually grow horns and become a highly fetishistic—and criminal—adult sexual identity. "He would type on a 3x5 card that he liked to squirt sperm into rubber caps or rubber girdles," wrote the detective, who in clichéd administrative dishevelment left the signature stain of a coffee mug on the police station memo. "Then [he would] place the cards in the victims' mail box and sometimes under the windshield wiper of their cars."

You may think this pathological rubber lover is an extreme case of sexuality gone awry, which it may very well be. But in studying the sexually abnormal, researchers can gain unique insight into the nuanced, otherwise hidden mechanisms of standard human sexual development and psychosexuality. The rubberphile's early childhood exposure to his mother's bathing suit, an impossibly white piece of material still beaded with lake water and fragrant with her perspiration, was perhaps simply coincident with a happenstance erection. Yet this chemistry was so powerful that once he massaged that elastic between his little thumb and his forefinger, all was forever lost.

This basic developmental system, one in which certain salient childhood events "imprint" on our developing sexualities, may not be terribly uncommon. In fact, that early childhood experiences mold our adult sexual preferences—specifically, what turns us on and off, however subtle or even unconscious these particular biases may be—could even be run-of-the-mill. And just like the institutionalized rubber lover, the more carnally humdrum among us might also owe our secret preferences in the bedroom to our becoming aroused, at some point in the distant past, by our own parents, relatives, or childhood friends.

Consider the case of a twenty-nine-year-old woman reported in an old *Archives of General Psychiatry* article who noticed to her dismay that she wasn't averse to a bit of sadomasochism and penis gazing when having sex with men. On accounting for these strong erotic triggers, the woman recalled:

> When I was four, my father once caught me masturbating. He put me over his knee and smacked my buttocks. He was in pajamas, and the slit in front of his trousers opened widely, so that I could see his big penis and dark scrotum moving quite near my mouth each time he raised his hand . . . Ever since, I subconsciously connected the smacking of my buttocks with the view of his penis and my first sexual excitement.

The trouble, of course, is that childhood sexual

experiences, and in particular their causal relationship with adult human sexuality, are an elusive topic to study, at least in any rigorously controlled sense. It's also an area of research that a prudish society—or at least one that views an individual's sexuality as appearing out of the blue with the first pubescent flush of hormones (or, alternatively, as unfolding in some highly innate, blueprinted sense impenetrable to experience, for example, "the gay gene")—prefers to look away from, in spite of its centrality to the human experience. Unlike, say, studying children's acquisition of language, examining the precise developmental pathways to adult sexuality is more or less impossible. That's not because it's empirically impossible but rather because childhood sexuality is one of those third-rail topics that gets zapped by the electric fencing of university ethics boards and is therefore at risk of always remaining little understood. So as intriguing as retrospective self-reports like the ones above may be, they are, alas, little more than anecdotes.

Yet never underestimate the cleverness of a good experimentalist. Although examining the precise causal links between early exposure to specific stimuli and adult sexuality is not exactly amenable to laboratory manipulation, there may be ways yet to explore general developmental mysteries using controlled methods. For example, for many investigative purposes, children are easily enough replaced by rats, and that's just what the researchers Thomas Fillion and Elliott Blass did in a now-classic study showing just how important early experiences can be in shaping adult sexual behavior. As reported in their 1986 study in *Science*, Fillion and Blass took three female rats that had just given birth to litters of pups and experimentally altered these mothers' odors in different ways. One of the rat dams had her teats and vagina coated with a lemony scent called citral; another dam had only her back coated with the same citral scent; and finally the third mother rat went lemonless—instead, her teats and vagina were simply painted with an odorless isotonic saline solution. So, if you'll follow this through, once the dams were placed back with their suckling babes, the litters of pups differed from one another with respect to the particular odor—or at least the location of the odor—emanating from their mothers while she nursed.

Once they were weaned, the young rats were removed permanently from their mothers and went about doing things that juvenile rats do. Then, at about a hundred days of age, the sexually mature male rats from these initial litters were introduced, individually, to one of two receptive female rats. Here's the trick, though. Prior to their introduction to the males, Fillion and Blass had coated one of these new females perivaginally with citral scent, and they left the other with her vagina smelling au naturel. Although the citral-scented female genitals made little difference to males from the two other litters—they were happy to have sex with either female—those males that, as pups, had suckled from a mother whose teats and vagina were redolent with lemon ejaculated significantly faster when they were now paired as adults with a lemony female sexual partner. In fact, the investigators reported that these males even had trouble achieving orgasm when mating with the odorless (or at least odorless as far as rat vaginas go) females.

But can we generalize these Oedipal rodent findings to the development of human sexuality? As far as I'm aware, similar studies have not been done with our own species— although it is interesting to speculate on the possible effects of human breastfeeding on the sexual preferences and biases of adult men. Tied as we are to the idea that children are asexual, however, it's unlikely we'll ever know for certain whether or not these data have any analogues with human sexuality; furthermore, I'd imagine it would be a real challenge to find mothers willing to tinker with their child's development in this domain. Turning one's son into a fetishist with an unhealthy attraction for reproductive organs that smell like lemon Pledge may well be going above and beyond the call of scientific duty, even if it is done for admirable reasons.

If only that long-forgotten Michigan rubberphile had known of such curious mechanisms of sexual imprinting, he might have found some solace in science rather than being relentlessly hounded by feelings of religious guilt. What an unfortunate thing to be the same as everyone else in underlying principle but, owing to something largely out of one's control, so different in technical expression.

Actually, perhaps it's not too late for him after all. In his letter, the detective wrote that our rubber lover was in the psychiatric ward, "where he hopes to spend the rest of his life and he wishes to live to be a real old man." According to my calculations, he'd be in his mid-eighties today. If the hospital staff was ever computer savvy and liberal minded enough to permit their inpatients to browse online, I do hope he lived long enough to experience the sexual renaissance of the Internet . . . he'd have found tens of thousands of others like him who would have happily indulged his fantasies without his having to resort to criminal activity. And maybe, just maybe, he's reading this book right now, thinking fondly of his mother clad in white rubber.

PART V

Ladies' Night

Female Ejaculation:
A Scientific Road Less Traveled

In spite of my own sexual biases, which I'll try to keep
from saturating our discussion, female ejaculation is an
enormously fascinating subject matter that has largely
escaped serious scientific inquiry, particularly from an
evolutionary perspective. This is all the more puzzling
given that female ejaculation, which is usually defined as
the expulsion of a *significant* amount of fluid around the
time of orgasm—estimates range from, on average, three
to fifty milliliters (about ten teaspoons)—is a topic that
was first described by scholars about two thousand years
ago. We're not talking here about the normal vaginal
lubrication that accompanies female arousal, but rather
something more akin to the copious seminal emissions that
occur with male orgasms.

In an extraordinary 2010 review article in *The Journal
of Sexual Medicine*, the urologist Joanna Korda and her
colleagues combed through the translated texts of the
ancient Eastern and Western literatures and plucked out
multiple references that would appear to distinguish
between common vaginal lubrication during intercourse
and the rarer external ejaculation of sexual fluids. The
fourth-century Taoist text *Secret Instructions Concerning*

the Jade Chamber, for example, written for the enterprising man on the art of satisfying a woman in bed, suggested that he decipher the following "five signs" of feminine arousal accordingly:

1. reddened face = she wants to make love with you
2. breasts hard and nose perspiring = she wants you to insert your penis
3. throat dry and saliva blocked = she is very stimulated and excited
4. slippery vagina = she wants to have her orgasm soon
5. the genitals transmit fluid = she has already been satisfied

I wouldn't recommend you implement these secret instructions today; citing number two in your defense that, say, some woman with a sweaty nose wanted you to insert your penis into her isn't likely to hold up in a court of law. But the fact that this ancient text distinguishes between "slippery vagina" and "the genitals transmit fluid," reason Korda and her coauthors, means that the latter can "clearly be interpreted as female ejaculation [at] orgasm." In ancient India, the *Kama Sutra*, which dates to A.D. 200–400, speaks of "female semen" that "falls continually." And in the West, even Aristotle had something to say about female discharge during sexual intercourse, which, he pointed out, "far exceeds" the seminal emission of the man. He also noted—and it's very tempting to speculate about just how he came to this conclusion—that female ejaculation tends to be "found in those who are fair-skinned and of a feminine type generally, but not in those who are dark and of masculine appearance."

It wasn't until the latter half of the seventeenth century, however, that the first truly scientific account of female ejaculation would be presented, this by a Dutch gynecologist named Reinier de Graaf, distinguishing precisely between vaginal lubrication, which accompanies arousal and facilitates intercourse, and female ejaculation, which is tantamount to seminal emission. "This liquid was clearly not designed by Nature to moisten the urethra (as some people think)," wrote de Graaf, describing the "pituitoserous juice" sometimes excreted around the time of female orgasm. "The ducts [from which they arise] are so placed at the outlet of the urethra that the liquid does not touch it as it rushes out."

Fast-forward to 1952, past the historical hordes of women secretly ejaculating in mass confusion, and we arrive at the offices of the German-born gynecologist Ernst Gräfenberg (curious how there were so many men in this profession), who, while the contributions of de Graaf and others are often overlooked, is credited with "discovering" an erotic zone on the anterior wall of the vagina running along the course of the urethra. Ernst, in other words, is the one who first christened the "G-spot" with his article "The Role of Urethra in Female Orgasm." In their review of his discovery, Korda and her colleagues report how Gräfenberg observed masturbating women (presumably in his office) expelling fluids from their urethra with orgasm "in gushes." Since this never occurred at the beginning of sexual stimulation, but rather only at the acme of orgasm, the physician concluded that its purpose was more for pleasure than for lubrication. "In the cases observed," wrote Gräfenberg, "the fluid was examined and it had no urinary character. I am inclined to believe that 'urine' reported to be expelled during female orgasm is not urine, but only secretions of the intraurethral glands correlated with the erotogenic zone along the urethra in the anterior vaginal wall."

It wasn't until 1982, in fact, that female ejaculate was first analyzed in terms of its chemical properties. If it's not urine, and it's not semen, then what, exactly, is it? After all, according to one study published by Amy Gilliland, most female ejaculators report "copious" amounts of fluid being released around the time of orgasm, enough to "soak the bed" or "spray the wall." So it's rather odd that we still don't have a name for this substance that at least 40 percent of women produce liberally at least once in their lives. (One especially clever reader suggested it be named "*ejill*culate," which I do like.)

Nearly all studies have shown a chemical dissimilarity between urine and female ejaculate—in fact, there are commonalities with male seminal fluid. You might recall from our previous discussion of male semen that only a small portion of that fluid contains sperm cells; the rest is a batter of psychotropic concoctions. Yet for many women, urine isn't entirely absent from the emission, either. Most female ejaculators, left to their own devices and without access to scientific information, describe their own explorations of the mysterious material. Some describe it as thick and viscous, or salty, others as watery and odorless. "No research has been done in this area for over 20 years," laments Gilliland, "and we still do not have an answer satisfying to most sexologists as to what female ejaculate fluid is or where it is manufactured."

Part of the trouble in investigating the phenomenon under properly controlled scientific conditions, however, is the fact that it doesn't particularly lend itself to laboratory investigations. According to most women, they need to be intensely aroused, as well as very relaxed, to ejaculate. So, although the clearest picture of what's happening down there would come from rigorous studies, the trouble is that subjecting self-reported female ejaculators to a barrage of invasive electromyographic laboratory techniques designed to stimulate their clitoris and evoke ejaculation kind of kills the mood. This is something that a team of Egyptian researchers learned the hard way. After attaching multiple electrodes to the genitals of thirty-eight healthy young women, as well as using vaginal and uterine balloons to measure pressure, and then stimulating the women to orgasm using electrovibration, they didn't find a drop of ejaculate, only vaginal lubrication. They could only surmise that foreplay might have done the trick. By contrast, although it involved another very small sample size, a team of Czechs did manage to evoke "female urethral expulsions" in ten women under laboratory conditions back in 1988, but these women, unlike those in the Egyptian study, had a self-reported history of frequent ejaculation.

In many ways, then, our best understanding to date of female ejaculation comes from the reports of female ejaculators themselves. But we do know from the chemical assays at least this: although it may have traces of urea, female ejaculate is obviously not urine. Many of the women interviewed by Gilliland recounted that after several humiliating episodes at this unexpected outburst of fluid, they'd since taken to voiding their bladders before having sex, yet still they ejaculated prodigiously. In fact, six of the thirteen women in the study had never even heard of female ejaculation prior to reading the study description; they just assumed they were "abnormal."

For most ejaculators, it doesn't happen every time an orgasm occurs. But this is in stark contrast to William Masters and Virginia Johnson's dubious 1966 assertion that female ejaculation is only an urban legend. Although some women were fortunate enough to find partners who enjoyed their ejaculations—partners would be right to assume, after all, that they're triumphant lovers if they can actually bring a woman to ejaculate—most had, at least at first, felt deep shame. In some cases, this translated to self-imposed celibacy and, not surprisingly, strained relationships. Education can change lives, even save marriages. One participant in Gilliland's study described the transformation in her husband after he understood that her ejaculation was a sign of her extraordinary sexual arousal: "Before he'd say, 'I don't want pee on me,' or 'Can't you go to the bathroom before sex?' Now he feels it's attractive and he'll say, 'Squirt me!' "

The good news is that many women note that they conceptualize their ejaculations in increasingly positive and empowering ways over the course of their lives. I'm very sympathetic to Gilliland's position when she concludes that "overall, it is the effect of ignorance about female ejaculation that should arouse us to action, not just scientific curiosity." I don't think that was an intentional pun on her part, by the way, but you do see how difficult it is to avoid them sometimes. Yet still, and please don't call me callous, I'm left enormously curious about the science. Why do only some women ejaculate and not others? What, if any, was the role of female ejaculation in human evolution? And why—just look at you now—is it such a giggle-inducing, fetishistic topic to some? Science has a long, wet, slippery challenge ahead indeed.

Studying the Elusive "Fag Hag":
Women Who Like Men Who Like Men

As a decades-long fan of *The Golden Girls*, I was saddened to learn of the death of Rue McClanahan in June 2010. In fact, I think I genuinely shed a palpable, detectable tear, which is something I can't remember ever doing on the death of a celebrity, with the exception perhaps of Bea Arthur and Estelle Getty. It sounds rather homosexually clichéd, I know, but my partner, Juan, and I have gotten into the habit of watching an episode of *The Golden Girls* every night before bed. And along with the other "girls," as we call them, McClanahan's character, Blanche Devereaux—the sassy southern belle with an insatiable appetite for rich cheesecake and rich men—has become something of an imaginary, smile-inducing friend in our home. Fortunately, Blanche's carnal spirit is burned forever on our DVDs. Yes, I know, I'm *so* gay.

The news of McClanahan's death inspired me to read more about her in real life—well, at least to expend enough finger energy to flitter over to her Wikipedia entry. I knew she'd been an outspoken advocate of gays and lesbians, as well as animals, but I didn't realize that her support for the former went all the way back to 1971. Just two short years after the Stonewall riots, she costarred in a movie set in a Greenwich gay bar called *Some of My Best Friends Are . . .* , and she just so happened to play a "vicious fag hag."

And then my mind switched gears, leaving the inimitable Rue and the issue of gay rights behind and instead focusing my attention on this term, *fag hag*. Now, I've never seen myself as a "fag"—although I'm sure many other people see me this way and unfortunately nothing more—but more important, I've certainly never regarded my many close female friends as "hags." So I was curious to learn more about the unflattering stereotypes lying at the etymological root of this moniker, which describes straight women who tend to gravitate toward gay men. Enter the psychologist Nancy Bartlett and her colleagues, who published the first quantitative study of "fag hags" in the journal *Body Image*.

These researchers, too, found the term intriguing. There are plenty of other colorful expressions that capture this distinct demographic rather vividly, some less insultingly so than others, including:

- Fruit fly
- Queen bee
- Queer dear
- Fairy godmother
- Fag shagger
- Queen magnet
- Hag along
- Swish dish
- Faggotina
- Homo honey
- Fairy collector
- Fairy princess
- Fagnet

But it's "fag hag" that resonates in the public conscious-ness. The researchers note that in both popular media and everyday expression, the term stirs up in most people's minds the image of an unattractive, overweight, desperate woman who seeks out the company of gay men to com-pensate for her lack of romantic attention from straight guys. Sorting through anecdotes from previous research, television, and cheap romance novels, the authors find that other common stereotypes paint the fag hag as being notoriously camp, overly emotional, unstable, and craving attention (think Megan Mullally's character Karen Walker from *Will & Grace*). What's especially fascinating is the authors' observation that this social category of women who like men who like men may be "cross-culturally robust": the French, they note, refer to such women as *soeurettes* (little sisters), the Germans brand them *Schwulenmuttis* (gay moms), and the Mexicans know them as *joteras* (*jota* is commonly used for "fag"). In Japan, these women are called *okoge*, translated literally as "the burned rice that sticks to the bottom of the pot."

According to the investigators, the "hag" component is essentially the common belief that these women "do not feel good about their bodies, and as a result, take refuge in the 'gay world' to avoid the harsher judgment and emphasis on female physical attractiveness inherent in the heterosexual social scene." The comedienne Margaret Cho, a well-known and self-proclaimed "fag hag," states: "The gay man in your life is not concerned with your youth and beauty. He wants to know your soul. He loves you for your courage and intellect. Whether you are lovely or plain, you are beautiful to him for these qualities—and many more."

As "the gay man" in many women's lives, I'm not sure Cho's got it entirely right about us; she seems to be idealizing gay men. There's certainly no shortage of vacuous, superficial gay men out there. She's also apparently never heard of the biologist John Maynard Smith's "sneaky fucker" evolutionary hypothesis for male homosexuality, which posits that gay men in the ancestral past had unique access to the reproductive niche because females let their guards down around them and other males didn't view them as sexual competitors. We're not infertile, after all, just gay. And stranger things have happened—especially when you toss some gin into the mix.

But what Bartlett and her coauthors were especially interested in was if there's any truth to the negative stereotypes surrounding fag hags. So they invited 154 heterosexual women to participate in an Internet-based survey on fag-haggery (my term, not theirs). These women ranged from seventeen to sixty-five years of age (with an average of twenty-eight years) and had a varied history of romantic relationships. Some were married, some single, still others divorced, widowed, currently dating, and so on, and most were reasonably well educated, having at least some college education. Each woman was asked to provide certain quantifiable information that would allow the authors to test several key hypotheses about myths surrounding the fag hag.

First, women simply gave the total numbers of gay male, straight male, and female friends they had. Also, out of these friends, they were asked to rate their degree of "closeness" with their best gay male, straight male, and female friend. Next, the women completed a commonly used instrument called the Body Esteem Scale, a twenty-four-item questionnaire measuring a woman's self-perceived sexual attractiveness and her weight concerns. Finally, each of the participants provided information about her romantic history over the past two years, including whether she'd been the "dumper" or the "dumpee" in recent failed relationships.

The results were analyzed to test the common assumption that women befriend gay men because they have poor body esteem and feel unattractive to straight men. If this was true, the authors reasoned, then there should be a meaningful statistical association between a woman's number of gay male friends and her body esteem and relationship success; in other words, the more pathetic a woman's romantic life and the more she sees herself as being undesirable to straight men, the more she should seek out gay men as friends. But the data revealed otherwise. In fact, with this sample at least, there was absolutely no link between a woman's relationship status, the number of times she'd been on the receiving end of a breakup, or her body esteem and the number of gay male friends in her life.

Debunking common assumptions in science is nothing new, and that goes for myths about the fag hag too. But there were also some unexpected findings in this study. For example, the more gay male friends that a woman had, the more sexually attractive she found herself. Now, obviously, this is a correlation, so we can only speculate on causality. It could be—as the authors suggest—that women with more gay male friends actually are physically more attractive than those with fewer gay male friends. Perhaps being around gay men offers these women some relief from the constant sexual overtures of straight men. (Bartlett's study measured only perceived self-attractiveness, not others' ratings of attractiveness, so this is an open question.) Alternatively, being surrounded by flattering gay men may elevate the fag hag's self-esteem, and because this attention comes from men, it may be especially effective in doing so. Interestingly enough, however, the longer that a woman has been friends with her closest gay male friend, the lower her self-perceived sexual attractiveness. Interpreting this unexpected finding, the authors suggest that it may actually reflect some core but nuanced truth of the "fag hag" stereotype: "Perhaps women who perceive themselves as less sexually attractive develop closer relationships with gay men." Others just go for superficial, less enduring attachments with them.

To my own favorite fairy princess, Ginger: This one's for you. I love you. For the rest of you, here's a final thought to scratch your heads over. It occurred to me while writing this essay that the social category of straight men who like to socialize with lesbians is astonishingly vacant in our society. Sure, you may hear about some random "dyke tyke" or "lesbro" (two terms that, unlike "fag hag," are hardly part of the popular slang vocabulary and actually required me to do some intensive Googling), but their existence is clearly minimal. Your guess is as good as mine as to why there's such a discrepancy in frequency between the two mixed-gender homo friendships.

Darwin's Mystery Theater Presents . . .
The Case of the Female Orgasm

I've written at some length about the curious evolution of the male reproductive system in our species, so it's only right to devote some time to the natural origins of a biological mechanism that doesn't involve the Y chromosome. (Well, at least it doesn't have to.) Needless to say, the subject of female orgasms isn't exactly my cup of tea. Being a gay man, I've always thought of the female orgasm as rather exotic and foreign, sort of like decorative basket weaving in a small African village. I could be wrong about this, but as far as I know, I've never even been in the same room as a woman having an orgasm, let alone given a woman one. So with that in mind, let's have a look at what's happening with those whose orgasmic bliss isn't neurologically grounded in something that protrudes seven inches (or so) from the rest of their bodies.

Fortunately, a handful of dedicated researchers have spent a lot more time on this issue than I have. Yet it's fair to say that even these scientists are still scratching their heads over the evolution of the female orgasm. Before we get into the nitty-gritty, let's make sure we're all on the same page about what a female orgasm actually is. A good working definition can be found in a report in the *Annual*

Review of Sex Research. According to the psychologist Cindy Meston and her colleagues:

> Female orgasm is a variable, transient peak sensation of intense pleasure, creating an altered state of consciousness, usually with an initiation accompanied by involuntary, rhythmic contractions of the pelvic striated circumvaginal musculature, often with concomitant uterine and anal contractions and myotonia that resolves the sexually induced vasocongestion (sometimes only partially), generally with an induction of well-being and contentment.

Actually, in light of that description and sans the female bits, perhaps it's not entirely foreign to me after all. In fact, in terms of evolutionary function, women having orgasms with men is almost as puzzling as men having orgasms with men. How many of us human beings were conceived in the wake of our mothers having orgasms may never be known, but the same mystery doesn't surround our fathers' orgasms that day. Unlike men, women don't need to have an orgasm in order to propagate their genes.

So, from a biological perspective, the "adaptive function" of the female orgasm is still hotly contested. Some theorists, including the late and legendary Stephen Jay Gould, have claimed that it serves no purpose at all but is instead only a quirky, functionless by-product of the ejaculatory response in males. In one of his more provocative essays, "Male Nipples and Clitoral Ripples," Gould fleshed out an old argument first made by the anthropologist Donald Symons. In 1979, Symons noted that early in embryological development, males and females share the same basic body plan. As a serendipitous consequence of selection for male ejaculation (which in straight men serves obvious reproductive purposes), some of the shared connective tissue and nervous system pathways in females were "accidentally" shaped for pleasure by evolution, too, leading happily to the occasional orgasm in sexually mature females. The clitoris is essentially the female version of the penis, since the two derive from the same embryological substrate. This also explains why female orgasms are achieved more by clitoral than vaginal stimulation.

Lest you think the by-product hypothesis was propagandistic, cooked up in some musty faculty lounge by ivory-tower misogynists, note that for years the main advocate of this position has been a female philosopher of biology named Elisabeth Lloyd. In fact, it was Lloyd who had initially given Gould his lead on Symons's thinking on the subject and who would later write a book strongly endorsing the by-product hypothesis called *The Case of the Female Orgasm*. Lloyd's book was roasted by many evolutionary thinkers because of the not-so-subtle feminist undertones in her writing; basically, she argues that female carnal bliss has been liberated from the ugly realities of reproductive biology. Her position? Ladies, go out—or stay home alone, your choice—and enjoy yourselves, your sexuality is about more than just baby making. But over the years, other empirically minded detectives have been working on this case as well, and many have begun to question the by-product account, claiming instead that the evidence does indeed point to a possible adaptive function of female orgasm.

To help you play along in the role of orgasmic sleuth, here are a few suggestive clues that researchers in this area have been trying to piece together into a plausible evolutionary story:

Clue #1: Twin-based evidence shows that orgasm

frequency has a modest heritable component. Uncomfortable as it may be to think of your flush-faced grandmother moaning in ecstasy, there is a clear genetic contribution to female orgasm. Hereditary factors account for only a third of the population-level variance in female orgasm, however.

Clue #2: Most women report that they are more likely to experience an orgasm while masturbating than during sexual intercourse with a male partner, and importantly, such masturbatory orgasms do not always hinge on simulating penile-vaginal sex. However, as the evolutionary psychologist David Barash notes, "just because something (e.g., female orgasm) can be achieved in diverse ways (e.g., masturbation) does not argue against it having evolved because it is particularly adaptive in a specific, different context (e.g., heterosexual intercourse)."

Clue #3: Educated women are more likely to report having masturbatory orgasms—but are no more likely to experience coital orgasms than are less educated women. Religiosity is another social mediator: religious women tend to have less frequent orgasms than nonreligious ones (or at least they report having fewer).

Clue #4: Using self-report data collected from college-aged American females, researchers such as the psychologist Todd Shackelford and the biologist Randy Thornhill have uncovered a positive correlation between frequency of orgasm and the physical attractiveness of male partners, with attractiveness being measured by subjective ratings as well as by indexes of facial symmetry. Recall that in "genetic fitness" terms, attractiveness tends to correlate positively with health and overall genetic value.

Clue #5: There is some physiological evidence that female orgasm leads to the retention of more and/or better-quality sperm among a single ejaculate. I don't think I can put it any better than the psychologists Danielle Cohen and Jay Belsky: "During the female copulatory orgasm the cervix rhythmically dips into the semen pool, thereby increasing sperm retention (by about 5 percent) relative to intercourse without orgasm, along with the probability of conception." But as Lloyd points out, most references to these classic "data" on the "uterine upsuck" properties of female orgasm derive from a single participant and were part of an old study done back in 1970. Nevertheless, tellingly, a woman's "desire to conceive" leads to more frequent self-reported orgasms during sex, and female orgasms are also most likely to occur during the most fertile period of the menstrual cycle.

Clue #6: In a provocative study by the psychologists Thomas Pollet and Daniel Nettle, Chinese women who were dating or married to wealthy male partners reported having orgasms more frequently than women whose partners made less money. That is to say, male partner income correlated strongly and positively with female orgasm frequency, and this income effect panned out even after the authors controlled for (ruled out) a host of extraneous variables, including health, happiness, education, the woman's personal income, and "Westernization." In any event, if we were to employ Pollet and Nettle's theory to other species, women may not be the only females in the animal kingdom whose orgasms are linked to the status and wealth of their male sexual partners. Japanese macaque females display the "orgasmlike" clutching reaction more often when they're mating with high-status males. There's no data yet on whether or not they also bite their lower lips in the process.

Together, these findings seemingly vindicate Barash, a vocal critic of Lloyd who, in fact, has been arguing that female orgasm "is a signal whereby a female's body tells her brain that she is sexually engaged with a [socially] dominant individual." Pollet and Nettle speculate that female orgasm may be linked to male income because money (resources) is a reliable indicator of the male's long-term investing in offspring and it may also reflect desirable underlying genetic characteristics. In this light, female orgasm may serve an emotional bonding role, motivating sexual behavior—and hence conception—with high-status males. This is one way to interpret those data, of course, but you may have some other ideas of your own. High-status males typically have higher self-esteem than other men, for instance, which possibly translates to their being better, more secure lovers in the boudoir. In other words, it could be that the men's actual *behavior* in the bedroom matters more than their social capital or their net worth.

As you can see, the natural origins of female orgasm remain somewhat mysterious. Some of the findings and logic favor the by-product hypothesis, whereas recent data on male quality and orgasm frequency cast reasonable doubt on the "functionless" accounts. What's more, female orgasm is unfortunately one of those questions that do not easily lend themselves to controlled experimentation in the laboratory. One can't, of course, randomly assign women to have sex with males differing in status and attractiveness to see if they climax or not. There are many other important avenues left to explore too, including whether orgasms in lesbians, for example, are tied to similar partner attributes as those above, or whether there's a different pattern with orgasm among gay women altogether.

I do wish there were a climax to the story and that I might satisfy you, but unfortunately this one doesn't have a tidy ending. As we've seen, some of the greatest minds in modern evolutionary biology have put their heads to the pleasure-filled pudenda, with astonishingly little success (or at least agreement). So in the end, I'm afraid I must leave it to you, dear readers, to piece together a once-upon-a-time story of female orgasm featuring the clues you're left with.

The Bitch Evolved:
Why Are Girls So Cruel to Each Other?

Not long ago I was invited to give a brief talk to my nephew Gianni's first-grade class—nothing too deep, mind you—simply about what it was like living in a foreign place such as Belfast. The highlight of my presentation was the uproarious laughter that erupted when I mentioned that people on that side of the Atlantic refer to diapers as "nappies" and cookies as "biscuits." But one must play to the audience.

Now, my sister resides in a small town in central Ohio, so perhaps there's something about the Midwest that breeds especially endearing and affectionate six-year-olds, but I should be forgiven for momentarily siding with Rousseau that afternoon on his overly simplistic view that society corrupts and turns such naive, innocent cherubs into monstrous adults. To give an example, one little girl waved at me in so kind a manner that it seemed, in that instant, I was in the presence of a better species of humankind, one that naturally regards other people as benevolent curiosities and one for whom the contrivances of social etiquette haven't tarnished and brutally tamed genuine emotions.

What shattered this rose-tinted illusion of mine was the

knowledge that these diminutive figures giggling and sitting cross-legged on the carpet before me might also be viewed as incubating adolescents. Perhaps it's just me, but I'd swear the world knows not a more sadistic soul than an angry, angst-ridden, hormonally intoxicated teen. And in just a few years' time this little pigtailed girl may morph into an eye-rolling, gossiping, ostracizing, sarcastic, dismissive, cliquish ninth grader, embroiled in the classic cafeteria-style bitchery of adolescent politics.

If that strikes you as misogynistic, rest assured it's merely an empirical statement. (Rest assured, also, that I'm afraid I have much in common with this tactical style, and I have great respect for more refined Machiavellians, so I'm not throwing stones here.) In fact, over the past few decades, scholars from a variety of disciplines—including developmental psychology, evolutionary biology, and cultural anthropology—have noted a striking difference in the standard patterns of aggression between males and females of reproductive age. While teenage boys and young male adults are more prone to engage in direct physical aggression, including hitting, punching, and kicking, females, by comparison, exhibit pronounced social aggression.

Here's a prototypical example, taken from a study in the *International Journal of Adolescence and Youth*:

Jo is a fifteen-year-old girl. She is average at her high school work and she is involved in school tennis in summer and netball in winter. In the past, she was well accepted, having a close group of friends and getting along well with most of her peers. After a day off with illness, she returns to school to find that things have changed. She walks over to her usual group but when she tries to talk to any of them, their responses are abrupt and unfriendly. She tries to catch the eye of her friend, Brooke, but Brooke avoids her gaze. In first lesson, she sits in her usual seat only to find that Brooke is sitting with someone else. At recess time, she joins the group late but just in time to overhear one of the girls bitching about her.

In peer discussion groups with teenage girls in South Australia, researchers found that Jo's situation is incredibly common. And what's especially sad is that adult authority figures such as teachers and parents often miss such devastating acts of reputational violence because they're so subtle and often occur "in context"—that is, they're less conspicuous than the physical altercations of boys.

Let me attempt to preempt the obvious criticism by noting that this is not, of course, to say all teenage girls are catty—need I really point out the obvious, that many are of course wonderful, thoughtful, and mature people? Nor is it to say that teenage boys are never socially aggressive or that girls don't display physical violence. But the culturally recurrent findings of female social aggression, and the largely invariant age distribution where such behaviors and attitudes are especially prominent (flaring up between about age eleven and age seventeen in girls), suggest a strong psychological bent in the fairer sex that leads "naturally" to these types of displays.

The anthropologists Nicole Hess and Edward Hagen explored this question of whether female social aggression is innate. They rounded up 255 undergraduates—men and women ranging from eighteen to twenty-five years of age—and asked them to read and mull over the following social scenario, which I'll summarize here.

Let's say you're at a campus party and out of the corner

of your eye you notice one of your classmates (another male student for male participants and another female student for female participants) conversing with the teaching assistant for a class you share with this other student. The other student is overheard saying some rather nasty lies about you; in particular, he or she is telling the teaching assistant that you haven't been working on a joint project for the class. Instead, this person says, you've been slacking off, coming to class with a hangover, and partying in Baja. Your TA glances over at you, with your beer in hand, and then glances away quickly as if disgusted. Then your duplicitous classmate walks over to you and says innocently, "Hey! How are things going? Hasn't the weather been great lately?"

Once participants read over this little story, they completed a questionnaire about how they would have liked to respond to this tattling person. On a scale of 1 to 10, with 1 being "disagree strongly" and 10 being "agree strongly," participants were asked to respond to statements such as "I feel like punching this person right now," "I feel like telling people at the party that this person is clueless and spews useless comments during lecture," and "I feel like saying, 'Yeah, the weather has been nice.' " Whereas the first two items are measures of direct and indirect aggression, respectively, the last item presumably tapped into the participant's willingness to turn the other cheek, so to speak. Importantly, Hess and Hagen also asked the participants how appropriate they thought various acts of violence against the treacherous classmate would be.

Their findings indicated a clear sex difference in aggressive responses, with women overwhelmingly compelled to retaliate by attacking the offender's reputation, mostly through gossip. This gender effect panned out even after controlling for participants' evaluation of the social appropriateness of such acts. In other words, even though the women realized malicious gossip wasn't socially appropriate, this was nevertheless their preferred first point of attack. Men, on the other hand, were more evenly divided in their response but failed to show the same preferential bias for acts of "informational warfare" against the unlikable classmate.

Although most researchers acknowledge the speculative nature of evolutionary arguments in this area, social aggression among reproductively viable females is usually interpreted as a form of mate competition. Hess and Hagen, for example, suggest that the sex differences uncovered in their study would likely have been even more pronounced in a younger group of participants. Evolutionarily, historically, and cross-culturally, they point out, girls in the fifteen- to nineteen-year-old range would be most actively competing for mates. Thus, anything that would sabotage another female's image as a desirable reproductive partner, such as commenting on her promiscuity, physical appearance, or some other aberrant or quirky traits, tends to be the stuff of virulent gossip.

Also, the degree of bitchiness should then demonstrate a sort of bell-shaped curve over the female life course. On the surface this seems mostly true. Anecdotally, I can't think of a single postmenopausal woman who seems hell-bent on undermining another woman's dating life—unless, perhaps, that involves spreading rumors about the sexual rival of her fertile daughter, in whom she has a vested adaptive interest. Then I can actually name names.

The psychologist Anne Campbell's work on sex differences and aggression has managed to carefully tease apart the many complex strands of cultural transmission and hormonal mediators in female violence. Campbell has argued that much of the sex differences in aggression can be understood in terms of "parental investment theory." Parental investment theory was developed in the early 1970s by the biologist Robert Trivers. One of its basic implications is that since human mothers make a disproportionately greater contribution to (and physical investment in) the offspring's survival than human fathers, women have evolved to be generally more reserved than men in mating strategies. Typical male physical violence, Campbell argues, is largely a form of showy sexual competition between men for reproductive access to the most desirable women. The type of social aggression we've just observed in women also appears to be a form of intrasexual competition for the most desirable men, but it avoids the comparatively higher cost of physical harm to women's precariously fertile bodies.

No parent wants to think he or she is rearing a socially insensitive daughter. But remember that psychological science is a discipline based on statistically significant, aggregate differences between comparison groups. In the present case, there are observed differences in the preferred aggressive-retaliatory styles between the sexes—ones that continue to appear even after controlling for social norms. But there's also, of course, fairly dramatic individual variation. The more we understand about the evolved pressures underlying our behaviors, the more we can get a grip on them and evaluate our own motives. One of my favorite thinkers, the feminist cultural constructivist Simone de Beauvoir, wrote famously that "one isn't born a woman, but becomes one." While it's true that culture exerts strong pressures shaping expressions of gender disparities, it also helps to know the biological mold that society must contend with.

PART VI

The Gayer Science:
There's Something Queer Here

Never Ask a Gay Man for Directions

I always seem to be the guy people ask for directions. That is to say, me, the spatially challenged, head-to-the-ground, asocial personality trying to avoid eye contact with any given passerby. This was especially the case while I was an expatriate living in Belfast. Usually I tried to wing it so that I didn't come across as completely stupid. But try as I might, my response always ended up sputtering its way into a wan shrug and the trusty fallback "I'm so sorry, but I'm an American. I'm afraid you've asked the wrong person." Given America's cartoon-character status throughout much of Europe, being an apologetically naive Yankee greased my way out of a lot of awkward social encounters in the United Kingdom, so this tactic usually worked just fine. (Unless I got a chatty person who wasn't in any hurry and I was his or her first live link to the New World. Then I was in for a lengthy discussion about Obama and Disney World.)

But the truth is, I called Northern Ireland home for almost six years and I should have been able to give directions like a local. It's not as if people were asking me how to get to some littleknown footpath in the Mourne Mountains; they wanted to know how to get to the pharmacy or the quickest route to the student union at the university where I worked. It's not just giving directions I struggle with, either. For as long as I can remember, I've had a knack for getting lost. I've wasted more of my life wandering around parking lots, hospitals, and campuses than I care to know. Maps? Anathema. I might as well be looking at Mayan hieroglyphics on a tree-bark roll.

What makes my "condition" even more ironic is that, according to family legend, I'm descended from the great Danish navigator Vitus Bering. Well, he wasn't all that great, since he got shipwrecked on the Commander Islands and lost nearly half his crew before dying of an unknown disease. But I imagine he would have at least needed to know his way around a nautical map to have been commissioned by Peter the Great and hailed as the first European to spy the southern shores of Alaska. So if I come from such Euclidean-headed genetic stock, why is my own brain slow as molasses when it comes to finding my way around town?

According to mounting evidence being gathered by the psychologist Qazi Rahman and his colleagues, it may well have something to do with the fact that I'm gay. Mind you, it's not that I'm poor at directions *because* I'm gay, but rather Rahman has discovered a nontrivial neural correlation between these two psychological traits. This correlation is similar in nature to the finding that left-handed individuals demonstrate better memory for events than right-handers due to their generally larger corpora callosa, a neurological boon that facilitates episodic recall. Southpaws are better at recalling memories not because they're left-handed but because of the common physical (brain) denominator underlying the expression of both traits.

Because of atypical hormonal influences on the developing fetus during prenatal growth, including the amount of circulating androgens (for example, testosterone) present in the mother's womb, homosexuals (both men and women) often display several telltale "bio-demographic" markers—residual bodily characteristics that indicate the prenatal effect of these hormonal factors. For example, you may already know about the well-publicized 2D:4D effect, scientific shorthand for the peculiar finding that for both straight women and gay men, the length ratio between the second and fourth digits (fingers) is, on average, greater than it is for gay women and straight men. Since the brain is just another physical template, there are also differences between straights and gays in brain structure (notably in the hippocampus) and therefore cognitive abilities. For example, gay men and straight women tend to outperform gay women and straight men on most verbal measures, whereas straight men out-perform the other groups on measures of spatial intelligence.

In a study reported in *Behavioral Neuroscience*, Rahman and his colleagues found that gay men are more like women than like straight men in that they are more dependent on left/right landmark strategies for navigation (for example, "turn right at the church") than on the Euclidean orientation strategies preferred by straight men (for example, "the bar is five miles in an easterly direction"). And in a follow-up study in the journal *Hippocampus*, Rahman and his coauthor, the psychologist Johanna Koerting, found that heterosexual males are different from gay men, straight women, and gay women in that they perform significantly faster on a task requiring them to scout out novel terrain in order to find a hidden search target. (Note that the researchers only tested people who regarded themselves as exclusively heterosexual or homosexual. Bisexuals were excluded.)

Now, before you go thinking up exceptions to these general findings, you contrarian you, note that they refer to aggregate population-level differences. Although I personally match Rahman's cross-sex neurocognitive model for gay brains to a tee, my partner Juan's brain is a satellite navigation device that could have given old Uncle Vitus's a run for its money. And Juan, unlike me, has a pronounced 2D:4D ratio. Furthermore, in science, a statistically significant difference between comparison groups may actually translate to negligible differences in the real world. Finally, Rahman is quick to point out that it's not as though gay men simply have women's brains, or that gay women have men's brains. Rather, the brains of homosexuals are more like neurocognitive mosaics of both sexes. For example, lesbians do not appear to differ from heterosexual women on cognitive measures except for verbal fluency, where they score in the male-typical direction.

A final note. I once happened upon a finding indicative of another physiological difference between homosexuals and heterosexuals. In addition to our navigational shortcomings, evidence suggests that gay people produce different armpit odors from straight people and that these scents are detectable. So perhaps if I stopped wearing deodorant, this would deter people from asking me for directions . . . along with just about anything else.

"Single, Angry, Straight Male
. . . *Seeks Same*":
Homophobia as Repressed Desire

I wish I could say that I decided to come out of the closet in my early twenties for more admirable reasons—such as love or the principle of the thing. But the truth is that passing for a straight person had become more of a hassle than I figured it was worth. Since the third grade, I'd spent too many valuable cognitive resources concocting deceptive schemes to cover up the fact that I was gay.

In fact, my earliest conscious tactic to hide my homosexuality involved being outlandishly homophobic. When I was eight years old, I figured that if I used the word "fag" a lot and on every possible occasion expressed my repugnance for gay people, others would obviously think I was straight. Although it sounded good in theory, I wasn't very hostile by temperament, and I had trouble channeling my fictitious outrage into convincing practice.

I may have failed as a homophobe, but unfortunately many people succeed. And it turns out we may have something in common: many young, homophobic males may secretly harbor homosexual desires (whether they are consciously trying to deceive the world about them, as I was, or are not even aware they exist). One of the most important lines of work in this area dates back to an article published in 1996 in the *Journal of Abnormal Psychology* in which the researchers Henry Adams, Lester Wright Jr., and Bethany Lohr report evidence that homophobic young males may secretly have gay urges.

In this study, sixty-four self-reported straight males with a mean age of twenty years were divided into two groups ("nonhomophobic men" and "homophobic men") on the basis of their scores on a questionnaire measure of aversion to gay males. Here, homophobia was operationally defined as the degree of "dread" experienced when placed in close quarters with a homosexual—basically, how comfortable or uncomfortable the person was interacting with gay people. (There is debate in the clinical literature about the semantics of this term, with some scholars introducing other constructs such as "homonegativism" to underscore the more cognitive nature of some people's antigay stance.)

Each participant then agreed to attach a penile plethysmograph to his, well, "lesser self." This device, which we've met before, is "a mercury-in-rubber circumferential strain gauge used to measure erectile responses to sexual stimuli. When attached, changes in the circumference of the penis cause changes in the electrical resistance of the mercury column." Previous research with this apparatus (the plethysmograph, not the penis—well, actually both) confirmed that significant changes in circumference occur only during sexual stimulation and sleep.

Next, the participants were led to a private chamber where they were shown three brief segments of graphic pornography. The three video snippets represented straight porn (scenes of fellatio and vaginal intercourse), lesbian porn (scenes of cunnilingus or "tribadism," which is, essentially, vulva rubbing), and gay male porn (scenes of fellatio and anal intercourse). Following each randomly ordered video presentation, the participant rated how sexually aroused he felt and also his degree of penile erection. Go on. *Guess the results.*

Both groups—nonhomophobic and homophobic men— showed significant engorgement to the straight and lesbian porn, and their subjective ratings of arousal matched their penile plethysmograph measure for these two types of videos. However, as predicted, only the homophobic men showed a significant increase in penile circumference in response to the gay male porn: specifically, 26 percent of these homophobic men showed "moderate tumescence" (six to twelve millimeters) to this video, and 54 percent showed "definite tumescence" (more than twelve milli- meters). (In contrast, for the nonhomophobic men, these percentages were 10 and 24, respectively.) Furthermore, the homophobic men significantly underestimated their degree of sexual arousal to the gay male porn.

From these data, the researchers concluded that "individuals who score high in the homophobic range and admit negative affect toward homosexuality demonstrate significant sexual arousal to male homosexual erotic stimuli." Of course, it isn't clear whether these people are unconsciously self-deceiving or consciously trying to con- ceal from others their secret attraction to members of the same sex. The Freudian defense mechanism of reaction formation—in which people's repressed desires are mani- fested by their fervent emotional reactions and hostile behaviors toward the very thing they desire—could explain the former. (From Shakespeare's *Hamlet*: "The lady doth protest too much, methinks.") The latter implies an act of deliberate social deception, such as my eight- year-old self's misguided scheming. It could of course be a bit of both, or work differently for different people. Who is to say whether all those inconveniently outed public figures (such as the televangelists Eddie Long and Ted Haggard, the conservative psychiatrist George Rekers, and the politicians Mark Foley and Larry Craig)—the very incarnations of this phenomenon—were self-deceiving or whether they knew they had full-blown homosexual urges all along?

Adams and his colleagues' interpretation of these plethysmograph findings have not gone unchallenged. In an article published in the *Journal of Research in Personality*, the researcher Brian Meier and his colleagues argue that Adams's findings can be better interpreted as the homophobic group's "defensive loathing" of gay males rather than a secret attraction. Drawing an analogy to other phobias, they state, "We believe it is inaccurate to argue that spider phobics secretly desire spiders or that claustrophobics secretly like to be crammed into dark and tight spaces." These investigators reason that Adams's homophobic sample experienced erections in response to the gay male porn due not to sexual arousal but to their anxiety over the images, which in turn provoked the physiological response of penile engorgement.

In my opinion, however, this "defensive loathing" re-interpretation by Meier is a bit off-kilter. Although it is true that ambient anxiety has been shown to increase the degree of sexual arousal in response to stimuli that are already sexually arousing, I could find no evidence that anxiety alone can give a man an erection. At least I hope this is the case. I get anxious about public speaking. If, on top of everything else, I have to worry about getting an erection during my talks, perhaps I ought to just cancel my appearances. Likewise, by these investigators' logic, male arachnophobes should get a mild tickle down there whenever they spy a spider scurrying across their desk. I suppose that's possible, but it seems rather far-fetched to me.

If we take Adams's findings that homophobic men get erections from watching gay porn as reasonable evidence of their sexual arousal, then these findings are enormously important. For example, they may help us to understand some of the psychological causes of gay bashing. Some of the most startling data I've come across involve a 1998 survey of five hundred straight men in the San Francisco area. Half of these men said they had acted aggressively in some way against homosexuals (and these were just the ones who admitted to such acts). And a third of those who hadn't struck out in this manner against gay people said that they would assault or harass a "homosexual who made a pass at them." If you missed the irony, this was in San Francisco—presumably one of the most "gay-friendly" places in the world.

In fact, a later study published in the *Journal of Abnormal Psychology* by Adams and his colleagues found that on a competitive task, homophobic men acted more aggressive toward gay male competitors than they did toward straight male competitors. In this study, fifty-two self-reported heterosexual men with a mean age of nineteen years were again classified as "homophobic" or "nonhomophobic" based on their responses to various items on a homophobia questionnaire. Participants were then told that they would be exposed to random types of erotic stimuli to determine pornography's effect on response time. In reality, all participants were shown only gay male porn.

Before and after watching this two-minute video of a male couple engaging in sexual foreplay, fellatio, and anal penetration, the participants completed several measures of their current emotional state (for example, whether they felt angry, anxious, sad, and so on). Then they proceeded to the competitive response-time task, where on twenty separate trials they were told to push a button as soon as a red "hit" light flashed on the console. Participants believed they were competing on this task against another player in an adjacent room. In fact, there was no other player, and the game was rigged so that on a randomly distributed half of the trials, the participant would lose. For every "winning" round, the participant was told he could deliver an electric shock varying in both degree and intensity to the other (nonexistent) player; alternatively, he had the option of administering no shock at all to this other person.

All players "lost" the first round and experienced a mild electric shock themselves, presumably administered by the other player. The critical manipulation in this study was that half of the participants thought they were competing against a gay male, whereas the other half thought they were competing against a straight male. Prior to the task, and after watching the gay porn, participants had been shown a brief video introducing them to this other "player." In one condition, this fictitious competitor was portrayed as a homosexual with stereotypical affectations who told the interviewer that he was in a "committed gay relationship with his partner, Steve, for two years." In the other condition, this same actor played it straight and said he was "involved in a committed dating relationship with his girlfriend for two years."

Although there was no significant difference between the homophobic and the nonhomophobic groups in the intensity and duration of shock administered to the straight competitor on winning trials, the homophobic group delivered more intense shocks and for longer durations when they thought the person in the other room was gay. On the subjective ratings of mood, the major difference between the two groups was on the dimension of anger-hostility: nonhomophobics showed a small positive blip in the radar on this dimension, while the homophobics showed a dramatic increase in anger-hostility between the pre-video measure of mood and the post-video rating. These data suggest that homoerotic stimuli—such as seeing two men holding hands—could send an already angry homophobic man over the top.

Although it is certainly true that the world today is more "approving" of homosexuality than it was just a decade ago—often begrudgingly so, in my opinion—there are still dangerous and malignant social elements beneath the surface preventing real acceptance. The day I can be in a public place in Anytown, U.S.A., and simply hold hands with the person I'm in love with (something most couples don't give a second thought to) without placing my partner and me in physical danger is the day I'll be convinced we've moved beyond rhetoric about "equal rights" and have actually changed hearts and minds.

Meanwhile, the next time you come across someone being especially hostile or reproachful toward gay people, stare him in the eye, scratch your chin, and repeat after me: "Hmm . . . very *interesting* . . ."

Baby-Mama Drama-less Sex: How Gay Heartbreak Rains on the Polyamory Parade

There's a strange whiff in the air, a sort of polyamory chic in which liberally minded journalists, an aggregate mass of antireligious pundits, and even scientists themselves have begun encouraging people to use evolutionary theory to revisit and revise their sexual attitudes and, more important, their behaviors in ways that fit their animal libidos more happily.

These recent attempts, including many bestselling books, explore how our modern, God-ridden, puritanical society conflicts with our species' evolutionary design, a tension making us pathologically ashamed of sex. There are of course many important caveats, but the basic logic is that because human beings are not naturally monogamous, but rather have been explicitly designed by natural selection to seek out "extra-pair copulatory partners"— having sex with someone other than your partner or spouse for the replicating sake of one's mindless genes— suppressing these deep mammalian instincts is futile and, worse, an inevitable death knell for an otherwise honest and healthy relationship.

Intellectually, I can get on board with this. If you believe, as I do, that we live in a natural rather than a supernatural world, then there is no inherent, divinely inspired reason to be sexually exclusive to one's partner. If you and your partner want to screw your neighbors on Wednesday nights after tacos, participate in beachside orgies lit by bonfires, or slip on your eyeless, kidskin discipline helmet and be led along by bridle and bit down the road to your local bondage society's weekly sex fest, then by all means do so (and take pictures). But the amoral beauty of Darwinian thinking is that it does not—or at least should not and cannot—prescribe any social behavior, sexual or otherwise, as being the "right" thing to do. Right is irrelevant. There is only what works and what doesn't work, within context, in biologically adaptive terms. And so even though any good and proper citizen is an evolutionarily informed sexual libertarian, Charles Darwin provides no more insight into a moral reality than, say, Dr. Laura Schlessinger.

On a related note, it's rather strange that we look for moral guidance about human sexuality from the rest of the animal kingdom, a logical fallacy in which what is "natural"—such as homosexual behavior in other species—is regarded as "acceptable." It's as if the fact that bonobos, desert toads, and emus have occasional same-sex liaisons has a moral bearing on gay rights in human beings. Even if we were the lone queer species in this god-less galaxy, even if it were entirely a "choice" between two consenting adults, why would that make it more reason-able to discriminate against people in homosexual relationships?

Beyond these philosophical problems with seeking out social prescriptions from a nature that is completely mute as to what we should do with our penises and vaginas, however, there's an even bigger hurdle to taking polyamory chic beyond the tabloids, talk shows, and

Internet forums and into standard bedroom practice. And that is simply the fact that we've evolved to empathize with other people's suffering, including the suffering of the people we'd betray by putting our affable genitals to their evolved promiscuous use.

Heartbreak is every bit as much a psychological adaptation as is the compulsion to have sex with those other than our partners, and it throws a monster of a monkey wrench into the evolutionists' otherwise practical polyamory. It's indeed natural for people—especially men, given, unlike women, their essentially unlimited reproductive potential—to seek sexual variety. My partner once likened this to having the same old meal over and over again, for years on end; eventually, you're going to get some serious cravings for a different dish. But I reminded him that people aren't the equivalent of a plate of spaghetti. Inconveniently enough, we have feelings.

Unless you have the unfortunate luck of being coupled with a psychopath, or have the good fortune of being one yourself, broken hearts are not easily experienced at either end, nor are they easily mended by reason or waved away by all the evolutionary logic in the world. And because we're designed by nature not only to be moderately promiscuous but also to become selfish when that natural promiscuity rears its head—again, naturally—in our partners, "reasonable people" are far from immune to getting hurt by their partner's open and agreed-upon sex with other parties. Monogamy may not be natural, but neither is indifference to our partners' sex lives or tolerance for polyamory. In fact, for many people, especially those naively taking guidance from scientists and pundits without thinking deeply enough about these issues, polyamory can lead to devastating effects.

One of the better accounts of the human heartbreak

experience is a summary by the anthropologist and author Helen Fisher. Drawing largely from work by psychiatrists, Fisher surmises that there are two main stages associated with a dead and dying romantic relationship, which is so often tied to one partner's infidelities. During the "protest" stage that occurs in the immediate aftermath of rejection, "abandoned lovers are generally dedicated to winning their sweetheart back. They obsessively dissect the relationship, trying to establish what went wrong; and they doggedly strategize about how to rekindle the romance. Disappointed lovers often make dramatic, humiliating, or even dangerous entrances into a beloved's home or place of work, then storm out, only to return and plead anew. They visit mutual haunts and shared friends. And they phone, e-mail, and write letters, pleading, accusing and/or trying to seduce their abandoner."

At the neurobiological level, the protest stage is characterized by unusually heightened, even frantic activity of dopamine and norepinephrine receptors in the brain, which has the effect of pronounced alertness similar to what is found in young animals abandoned by their mothers. This impassioned protest stage—if it proves unsuccessful in reestablishing the romantic relationship—slowly disintegrates into the second stage of heartbreak, what Fisher refers to as "resignation/despair," in which the rejected party gives up all hope of ever getting back together. "Drugged by sorrow," writes Fisher, "most cry, lie in bed, stare into space, drink too much, or hole up and watch TV." At the level of the brain, overtaxed dopamine-making cells begin sputtering out, causing lethargy and depression. And in the saddest cases, this depression is linked to heart attacks or strokes, so people can, quite literally, die of a broken heart. So we may not be "naturally monogamous" as a species, but neither are we entirely naturally polygamous.

It's depressing to even read about, I realize, but for most people those all-important chemicals eventually begin pulsating again when a new love affair begins. Let me note, however, that one of the more fascinating things about the resignation/despair stage is the possibility that it actually serves an adaptive function that may help to salvage the doomed relationship, especially for an empathetic species such as our own. As I mentioned earlier, heartbreak is not easily experienced at either end, and when your actions have produced such a sad and lamentable reaction in another person, when you watch someone you care about (but no longer feel any real long-term or sexual desire to be with) suffer in such ways, it can be difficult to fully extricate yourself from a withered romance. If I had to guess—in the absence of any studies that I'm aware of to support this claim—I'd say that a considerable amount of genes have replicated in our species solely because, with our damnable social cognitive abilities, we just don't have the heart to break other people's hearts.

Again, we may not be a sexually exclusive species, but we do form deep romantic attachments, and the emotional scaffolding on which these attachments are built is extraordinarily sensitive to our partners' sexual in-discretions. I also say this as a gay man who, according to mainstream evolutionary thinking, shouldn't be terribly concerned about his partner having sex with strangers. After all, it isn't as though he's going to get pregnant and cuckold me into raising another man's offspring. But if you'd explained that to me as I was screaming invectives at one of my partners following my discovery that he was cheating on me, curled up in the fetal position in the corner of my kitchen and rocking myself into self-pitying oblivion, or as I was vomiting my guts out over the toilet for much of the next two weeks, I would have nodded in rational Darwinian assent while still trembling like a wounded animal.

In fact, sexual jealousy in homosexual relationships, especially gay male relationships, is a place where polyamory chic meets with some significant theoretical problems. One of the most frequently cited findings in evolutionary psychology is the fact that men tend to become most jealous when their female partners have sex with other men, whereas women tend to get most jealous when their male partners show signs of "emotional infidelity" (behaviors that indicate that the man may be interested in "more than sex" with another woman and has developed meaningful feelings for her, possibly signaling long-term plans). These aren't mutually exclusive types of jealousy, mind you, but instead they represent points along a continuum or spectrum of jealousy—emotional jealousy at one end and sexual jealousy at the other. Males and females simply tend to fall, on average, at different places along the way in terms of what triggers their highest levels of jealousy. This general sex difference makes sense from an evolutionary perspective. Prior to the era of DNA testing, which is of course when human brains evolved, men were extremely vulnerable to investing, unwittingly, in some other guy's genes (conveniently packaged in the form of children). By contrast, women, encumbered by the many physical demands of bearing and tending to young children, would have evolved to rely primarily on their male long-term partner to help them raise their offspring to reproductive age. Therefore, they'd have been at risk of having his attention and resources diverted to another woman and her offspring.

So when it comes to homosexual affairs, writes the psychologist Brad Sagarin and his colleagues in *Evolution and Human Behavior*, "a same-sex infidelity does not entail the asymmetrical threats of mistaken paternity and of resources being diverted to another woman's children, suggesting both that the sexes may be similar in their jealous responses and that such responses may be less intense than in the case of opposite-sex infidelities." In fact, in studies designed to test this basic hypothesis, the researchers indeed found that jealousy was less intense when straight participants were asked how they would feel, hypothetically, if their partners had a homosexual fling than if they were to become involved with someone from the opposite sex. Personally, I think the participants would have other things to worry about besides jealousy if their partners were on the down low, but these data clearly show that reproduction-related concerns indeed moderate feelings of jealousy in human romantic relationships.

But the foregoing study actually highlights bisexual affairs, since the hypothetical cheating spouse is in a primary sexual relationship with someone of the opposite sex. By contrast, from the perspective of a same-sex partner in a long-term relationship, homosexual infidelity may elicit a different pattern of jealousy altogether. After all, as any gay person with a past knows, homosexual relationships certainly aren't without their fair share of this type of drama. Gay men may, in fact, be less dis-tressed by sexual infidelity than are straight men. But there are meaningful individual differences in this regard. Still, I'm willing to speculate and say this: most of us certainly aren't completely okay with the idea of our partners having sex with whomever they please. Nor, I'd imagine, are most lesbians comfortable with their partners seeing other lesbians and developing close relationships with them (that is, emotional infidelity). Now, perhaps I'm in a minority in caring so much about my partner's same-sex behaviors—at least

the ones not including me. When asked in a 2010 interview by a reporter for *New York* magazine how he'd feel
if his husband, Terry, cheated on him, the well-known sex
columnist Dan Savage, for example, said that he "[would-
n't] give a shit" and that gay men are "not psycho like
straight people are" about sexual infidelity in their partners. I'm not so sure about that. Often we're just as
psycho. In my case, I informed the sexual interloper that I
would gladly emasculate him with a crisp pair of scissors
if ever he made contact again with my partner. This was
classically aggressive "mate-guarding" behavior as seen in
straight men threatening their sexual rivals. Scaring off
other men like this, most evolutionary theorists believe, is
a preemptive tactic designed to thwart cuckoldry.

Gay men, of course, are unusually vulnerable to HIV,
and that's reason enough to become absolutely furious
about a partner's cheating behind one's back. Yet although
they're often commingled, jealousy is distinct from anger.
Also, the deadly scourge that is AIDS wasn't present in the
ancestral past, so the threat of this disease could not have
produced any special adaptive psychological defenses in
gay men's brains. So how else could one explain sexual
jealousy among gay men? It may actually be understood
by some sort of pseudo-heterosexuality mind-set, in which
gay men's brains are just the same as straight men's brains
in this regard—hypervigilant against being deceived into
raising some other man's child. All this is to say that I
reacted the way I did at my partner's cheating on me
because, at an unconscious level, I didn't want my
testiculared honey getting impregnated by another man. I
didn't consciously think of him as a woman, mind you; in
fact, if I had, I assure you I wouldn't have been with him.
But tell that to my gonads and my amygdala. I do wonder,
also, whether these differences may be related to whether

one is more a "top" or a "bottom," a subject we'll examine in the next essay.

—Signed, no longer woefully yours, yet in perpetually ready and melodramatic sorrow, your once, and likely future, heartbroken gay friend, J.B.

Top Scientists Get to the Bottom of Gay Male Sex Role Preferences

It's my impression that many straight people believe that there are two types of gay men in this world: those who like to give, and those who like to receive. No, I'm not referring to the relative generosity or gift-giving habits of homosexuals. Not exactly, anyway. Rather, the distinction concerns gay men's sexual role preferences when it comes to the act of anal intercourse. But like most aspects of human sexuality, it's not quite that simple.

I'm very much aware that some readers may think that this type of discussion isn't proper science. But the great thing about good science is that it's amoral and objective and doesn't cater to the court of public opinion. Data don't cringe; people do. Whether we're talking about a penis in a vagina or one in an anus, it's human behavior all the same. The ubiquity of homosexual behavior alone makes it fascinating. What's more, the study of self-labels in gay men has considerable applied value, such as its possible predictive capacity in tracking risky sexual behaviors and safe-sex practices.

People who derive more pleasure (or perhaps suffer less anxiety or discomfort) from acting as the "insertive" partner are referred to colloquially as *tops*, whereas those who have a clear preference for serving as the receptive partner are commonly known as *bottoms*. There are plenty of other descriptive slang terms for this gay male dichotomy as well, some repeatable ("pitchers versus catchers," "active versus passive," "dominant versus submissive") and others, well, not by a gentleman.

In fact, survey studies have found that many gay men actually self-identify as "versatile," which means that they have no strong preference for either the insertive or the receptive role. For a small minority, the distinction doesn't even apply, since some gay men lack any interest in anal sex and instead prefer different sexual activities. Still other men refuse to self-label as tops, bottoms, versatiles, or even gay at all, despite their having frequent anal sex with gay men. These are the so-called men who have sex with men (or MSM) who often have heterosexual relations as well and tend to see themselves as straight rather than bisexual.

Several years ago, a team of scientists led by Trevor Hart at the Centers for Disease Control and Prevention studied a group of 205 gay male participants. Among the group's major findings were these:

1. Self-labels are meaningfully correlated with actual sexual behaviors. That is to say, based on self-reports of their recent sexual histories, those who identify as tops are indeed more likely to act as the insertive partner, bottoms are more likely to be the receptive partner, and versatiles occupy an intermediate status in sex behavior.

2. Compared with bottoms, tops are more frequently engaged in (or at least they acknowledge being attracted to) other insertive sexual behaviors. For example, tops also tend to be the more frequent

insertive partner during oral intercourse. In fact, this finding of the generalizability of top/bottom self-labels to other types of sexual practices was also uncovered in a study showing that tops were more likely to be the insertive partner in everything from sex-toy play to verbal abuse to urination play (aka "water sports").

3. Tops were more likely than both bottoms and versatiles to reject a gay self-identity and to have had sex with a woman in the past three months. They also manifested higher internalized homophobia— essentially the degree of self-loathing linked to their homosexual desires.

4. Versatiles seem to enjoy better psychological health. Hart and his coauthors speculate that this may be due to their greater sexual sensation seeking, lower erotophobia (fear of sex), and greater comfort with a variety of roles and activities.

One of the primary aims of this study was to determine if self-labels in gay men might shed light on the epidemic spread of the AIDS virus. In fact, self-labels failed to correlate with unprotected intercourse and thus couldn't be used as a reliable predictor of condom use. Yet the authors make an excellent—potentially lifesaving—point:

Although self-labels were not associated with un-protected intercourse, tops, who engaged in a greater proportion of insertive anal sex than other groups, were also less likely to identify as gay. Non-gay-identified MSM [again, men who have sex with men] may have less contact with HIV-prevention messages and may be less likely to be reached by HIV-prevention programs than are gay-identified men. Tops may be less likely to be recruited in venues frequented by gay men, and their greater internalized homophobia may result in greater denial of ever engaging in sex with other men. Tops also may be more likely to transmit HIV to women because of their greater likelihood of being behaviorally bisexual.

Beyond these important health implications of the top/bottom/versatile self-labels are a variety of other personality, social, and physical correlates. Some psychologists point out that prospective gay male couples might want to weigh this issue of sex role preferences seriously before committing to anything long-term. From a sexual point of view, there are obvious logistical problems of two tops or two bottoms being in a monogamous relationship. But since these sexual role preferences tend to reflect other behavioral traits (such as tops being more aggressive and assertive than bottoms), "such relationships also might be more likely to encounter conflict quicker than relationships between complementary self-labels."

Another intriguing study was reported in the *Archives of Sexual Behavior* by the anthropologist Matthew McIntyre. McIntyre had forty-four gay male members of Harvard University's gay and lesbian alumni group mail him clear photocopies of their right hand along with a completed questionnaire on their occupations, sexual roles, and other measures of interest. This procedure allowed him to investigate possible correlations between such variables and the well-known 2D:4D effect, which I mentioned in my essay on gay men and navigational skills. Somewhat curiously, McIntyre discovered a small but statistically significant negative correlation between 2D:4D and sexual self-label. That is to say, at least in this small sample of gay Harvard alumni, those with the more masculinized 2D:4D profile were in fact more likely to report being on the receiving end of anal intercourse and to demonstrate more "feminine" attitudes in general.

Many questions about gay self-labels and their relation to development, social behavior, genes, and neurological substrates remain to be answered; indeed, they remain to be asked. That many gay men go one step further and use secondary self-labels, such as "service top" and "power bottom" (a pairing in which the top is actually submissive to the bottom), reveals even further complexity. For the right scientist, there's a lifetime of hard work just waiting to be had.

Is Your Child a "Pre-homosexual"? Forecasting Adult Sexual Orientation

There are signs, some would say omens, glimmering in certain children's demeanors that, probably ever since there *were* children, have caused parents' brows to crinkle with worry, precipitated forced conversations with nosy mothers-in-law, strained marriages, and ushered untold numbers into the deep covenant of sexual denial. We all know the stereotypes: an unusually light, delicate, effeminate air in a little boy's step, often coupled with solitary bookishness, or a limp wrist, an interest in dolls, makeup, princesses, dresses, and a strong distaste for rough play with other boys; in little girls, there is the outwardly boyish stance, perhaps a penchant for tools, a lumbering gait, a square-jawed readiness for physical tussles with boys, an aversion to all the perfumed, delicate, laced trappings of femininity.

So let's get down to brass tacks. It's what these behaviors signal to parents about their child's incipient sexuality that makes them so undesirable; these behavioral patterns are feared, loathed, and often spoken of directly as harbingers of adult homosexuality. However, it is only relatively recently that developmental scientists have conducted controlled studies with one clear aim in mind: to accurately identify the earliest and most reliable signs of adult homosexuality. In looking carefully at the childhoods of gay adults, researchers are finding an intriguing set of behavioral indicators that homosexuals seem to have in common. And, curiously enough, the age-old homophobic fears of many parents reflect some genuine predictive currency.

In their technical writings, researchers in this area simply refer to pint-sized prospective gays and lesbians as "pre-homosexual." This term isn't perfect: it manages to achieve an uncomfortable air of biological determinism and clinical interventionism simultaneously. But it is, at least, probably fairly accurate. Although not the first scientists to investigate the earliest antecedents of same-sex attraction, J. Michael Bailey, a psychologist, and the psychiatrist Kenneth Zucker published a seminal paper on childhood markers of homosexuality with their controversial article in *Developmental Psychology* in 1995. The explicit aim of this paper, according to the authors, "was to review the evidence concerning the possible association between childhood sex-typed behavior and adult sexual orientation." So one thing to keep in mind is that this particular work isn't about identifying the causes of homosexuality per se but instead about indexing the childhood correlates of same-sex attraction. In other words, nobody is disputing the likely genetic factors underlying adult homosexuality or the well-established prenatal influences. Instead, it is simply meant to index the nonerotic behavioral clues that best predict which children are most likely to be attracted, as adults, to those of the same sex and which are not.

By "sex-typed behavior," Bailey and Zucker are referring to that long, now scientifically canonical list of innate sex differences in the behaviors of young males versus young females. In innumerable studies, scientists have documented that these sex differences are largely impervious to learning and found in every culture examined (even, some researchers believe, in youngsters of other primate species). Now, before that argumentative streak in you starts whipping up exceptions to the rule— *obviously, there is variance both among and within individual children*—I hasten to add that it's only when comparing the aggregate data that sex differences leap into the stratosphere of statistical significance. The most salient among these differences are observed in the domain of play. Boys engage in what developmental psychologists refer to as "rough-and-tumble play," which is pretty much exactly what it sounds like, whereas girls prefer the company of dolls to a knee in the ribs.

In fact, toy interests are another key sex difference, with boys gravitating toward toy machine guns and monster trucks, and girls orienting toward baby dolls and hyperfeminized figurines. Young children of both sexes enjoy fantasy—or pretend—play, but the roles that the two sexes take on within the fantasy context are already clearly gender segregated by as early as two years of age, with girls enacting the role of, say, cooing mothers, ballerinas, or fairy princesses and boys strongly preferring more masculine characters, such as soldiers and superheroes. Not surprisingly, therefore, boys naturally select other boys for playmates, and girls would much rather play with other girls than with boys.

So on the basis of some earlier, shakier research, along with a good dose of common sense, Bailey and Zucker hypothesized that homosexuals would show an inverted pattern of sex-typed childhood behaviors (little boys preferring girls as playmates and infatuated with their mothers' makeup kits; little girls strangely enamored with field hockey or professional wrestling . . . that sort of thing). Empirically, explain the authors, there are two ways to investigate the relation between sex-typed behaviors and later sexual orientation. The first of these is to use a prospective method, in which young children displaying sex-atypical patterns are followed longitudinally into adolescence and early adulthood, such that the individual's sexual orientation can be assessed at reproductive maturity. Usually this is done by using something like the famous Kinsey scale, which involves a semistructured clinical interview about sexual behavior and sexual fantasies to rate people on a scale of 0 (exclusively heterosexual) to 6 (exclusively homosexual). I'm a solid 6; like Stephen Fry I wanted to get out of a vagina at one point in my life, but ever since then I've never had the slightest interest in going back into one.

Conducting prospective studies of this sort is not terribly practical, explain Bailey and Zucker, for several reasons. First, given that a relatively small proportion of the overall population is exclusively homosexual, a rather large number of prehomosexuals are needed to obtain a sufficient sample size, and this would require a huge oversampling of children *just in case* a small subset turns out gay. Second, a longitudinal study tracking the sexuality of children into late adolescence takes time—around sixteen years—so the prospective approach is very slow going. Finally, and perhaps the biggest problem with prospective homosexuality studies, not a lot of parents are likely to volunteer their children. Right or wrong, this is a sensitive topic, and usually it's only children who present significant sex-atypical behaviors—such as those with gender identity disorder—who are brought into clinics and whose cases are made available to researchers.

For example, the psychologist Kelley Drummond and

her colleagues interviewed twenty-five adult women who were referred by their parents for assessment at a mental health clinic when they were between three and twelve years of age. At the time, all of these girls had several diagnostic indicators of gender identity disorder. They might have strongly preferred male playmates, insisted on wearing boys' clothing, favored rough-and-tumble play over dolls and dress-up, stated that they would eventually grow a penis, or refused to urinate in a sitting position. As adults, however, only 12 percent of these women grew up to have gender dysphoria (the uncomfortable sense that one's biological sex does not match one's gender identity). Rather, the women's childhood histories were much more predictive of their adult sexual orientation. In fact, the researchers found that the odds of these women reporting a bisexual/homosexual orientation were up to twenty-three times higher than would normally occur in a general sample of young women. Not all tomboys become lesbians, of course, but these data do suggest that lesbians often have a history of cross-sex-typed behaviors.

And the same holds for gay men, according to Bailey and Zucker. They revealed that in retrospective studies (the second method used to examine the relation between childhood behavior and adult sexual orientation, in which adults simply answer questions about their childhoods), 89 percent of randomly sampled gay men recalled cross-sex-typed childhood behaviors exceeding the heterosexual median. Some critics have questioned the general retrospective approach, arguing that participants' memories (those of both gay and straight individuals) may be distorted to fit with societal expectations and stereotypes about what gays and straights are like as children. But in a rather clever study published in *Developmental Psychology*, evidence from childhood home videos validated the retrospective method by having people blindly code child targets on the latter's sextypical behaviors, as shown on the screen. The authors found that "those targets who, as adults, identified themselves as homosexual were judged to be gender nonconforming as children." Numerous studies have since replicated this general pattern of findings, all revealing a strong link between childhood deviations from gender role norms and adult sexual orientation. There is also evidence of a "dosage effect": the more gender-nonconforming characteristics there are in childhood, the more likely it is that a homosexual/bisexual orientation will be present in adulthood.

But—and perhaps you've been waiting for me to say this—there are several very important caveats to this body of work. Although gender-atypical behavior in childhood is strongly correlated with adult homosexuality, it is still an imperfect correlation. Not all little boys who like to wear dresses grow up to be gay, nor do all little girls who despise dresses become lesbians. Many will be straight, and some, let's not forget, will be transsexuals. Speaking for myself, I was rather androgynous, showing a mosaic pattern of sex-typical and sex-atypical behaviors as a child. In spite of my parents' preferred theory that I was simply a young Casanova, Zucker and Bailey's findings may account for that old Polaroid snapshot in which eleven of the thirteen other children at my seventh birthday party are little girls. But I also wasn't an overly effeminate child, was never bullied as a "sissy," and, by the time I was ten, was indistinguishably as annoying, uncouth, and wired as my close male peers.

In fact, by age thirteen, I was deeply socialized into masculine norms. In this case, I took to middle school wrestling as a rather scrawny eighty-pound eighth-grader, and in so doing, I ironically became all too conscious indeed of my homosexual orientation. Cross-cultural data show, actually, that pre-homosexual boys are more attracted to solitary sports, such as swimming, cycling, and tennis, than they are to rougher contact sports, such as football and soccer; they're also less likely to be child-hood bullies. (Prospective gay males who adapt too rigidly to the perceived gender norms as they grow older may, in fact, become hypermasculinized to such a degree that, as we've seen already, they also become dangerously homo-phobic in the process.) In any event, I distinctly recall being with the girls on the monkey bars during second-grade recess while the boys were in the field playing football, and looking over at them thinking how it was rather strange that anyone would want to act that way.

Another caveat is that researchers in this area readily concede that there are likely multiple—and no doubt very complicated—developmental routes to adult homo-sexuality. Heritable, biological factors interact with environmental experiences to produce phenotypic out-comes, and this is no less true for sexual orientation than it is for any other within-population variable. Since the prospective and retrospective data discussed in the fore-going studies often reveal very early emerging traits in pre-homosexuals, however, those children who show pro-nounced sex-atypical behaviors may have "more" of a genetic loading to their homosexuality, whereas gay adults who were sex typical as children might trace their homo-sexuality more directly to particular childhood experiences. For example, in a rather stunning case of what we might call "say-it-isn't-so science"—science that produces data that rebel against popular, politically correct, or emotionally appealing sentiments—recent con-troversial findings in the *Archives of Sexual Behavior* hint that men—but not women—who were sexually abused as children are significantly more likely than nonabused males to have had homosexual relationships as adults. Whatever the causal route, however, none of this implies, whatsoever, that sexual orientation is a choice. In fact, it implies quite the opposite, since, as we know from the rubber lover and foot fetishists we met earlier in this book, prepubertal erotic experiences can later consolidate into irreversible sexual orientations and preferences.

It is fashionable these days, particularly in the West, to say that one is "born gay." I appreciate the anti-discriminatory motives and believe strongly that this attitude reflects an increasingly humanitarian ethos toward sexual minorities. But if we think about it more critically, it's exceedingly odd, and nonsensical, to refer to a newborn infant still dripping with amniotic fluid as being a member of the LGBT community. Yes, it requires a prodigious degree of stupidity to talk about what makes one's genitalia become tumescent as being a conscious choice, but it's far from obvious that every person shoots equally out of their mother's birth canal with an already discriminating taste for penises over vaginas, or vice versa.

Then we arrive at the most important question of all. Why do parents worry so much about whether their child may or may not be gay? You might not be one of these fretful parents; in fact, you might like to see yourself as being indifferent to your child's sexuality as long as he or she is happy. Then again, all else being equal, I suspect we'd be hard-pressed to find parents who would actually prefer their offspring to be homosexual rather than heterosexual. Evolutionarily, needless to say, parental homophobia is a no-brainer: gay sons and lesbian daughters aren't likely to reproduce (unless they get creative). And I would imagine, on a viable hunch, that even in today's most liberally minded communities, coming out of the closet to parents is a much easier thing to do for gay individuals who have the luxury of demonstrably straight siblings who can carry their own reproductive weight. As for me, with a breeding older brother and sister—not with each other, mind you—and their little respective litters of my fantastic nieces and nephews, my father at least doesn't have to worry about his genes going extinct. In any event, I think it's far better for parents to recognize the source of their concerns about having a gay child as being motivated by *unconscious genetic interests* than it is to have them fibbing to themselves about being entirely indifferent to their son or daughter "turning out" gay.

And, bear this in mind, parents, it's also important to stress that since genetic success is weighed in evolutionary biological terms as the relative percentage of one's genes that carry over into subsequent generations—rather than simply number of offspring per se—there are other, though typically less profitable, ways for your child to contribute to your overall genetic success than humdrum sexual reproduction. For example, I don't know how much money or residual fame is trickling down to, say, k. d. lang, Elton John, and Rachel Maddow's close relatives, but I can only imagine that these straight kin are far better off in terms of their own reproductive opportunities than they would be without a homosexual dangling so magnificently on their family trees. The very thought of making love to a blood relative of Michelangelo or Hart Crane, irrespective of anything else about that person save his heritage, makes me strangely and instantly aroused, and I'd imagine such a person would be eminently desirable to heterosexually fecund women as well. So here's my message: cultivate your little pre-homosexual's native talents, and your ultimate genetic payoff could, strangely enough, be even larger with one very special gay child than it would be if ten mediocre straight offspring leaped from your loins.

There's one final thing to note, and that's in reference to the future of this research and its real-world applications. If researchers eventually perfect the forecasting of adult sexual orientation in children, what are the implications? Should broadminded mothers be insouciantly describing their OshKosh B'Gosh-wearing toddlers as "bi" or fathers relaying how their "straight" daughters started eating solid food or took their first steps at the grocery store today? Would parents want to know? Parents often say to their gay children, in retrospect, "I knew it all along." But hindsight is twenty-twenty, and here we're talking about the possibility of really, definitively, no-doubt-about-it, *knowing* your child is going to be gay from a very, very early age.

I can say as a once-pre-homosexual pip-squeak that some preparation on the part of others would have made it easier on me, rather than my constantly fearing rejection or worrying about some careless slipup leading to my "exposure." It would have at least avoided all of those awkward, incessant questions during my teenage years about why I wasn't dating a nice pretty girl (or questions from the nice pretty girl about why I was dating her and rejecting her advances).

And another thing: it must be pretty hard to look into your pre-homosexual toddler's limpid eyes, brush away the cookie crumbs from her cheek, and toss her out on the streets for being gay.

PART VII

For the Bible Tells Me So

Good Christians (but Only on Sundays)

This is a difficult confession to make, because, on the surface, I'm sure it sounds wildly hypocritical. Still, here goes: I trust religious people more than I trust atheists. The hypocritical part is that I happen to be an atheist with unshakably strong godless convictions. In my book *The Belief Instinct*, I've tried to explain at considerable length, in fact, why I feel this particular way. But for our purposes here the only important thing to know is that I've not a sliver of agnostic hesitation in my belief that there is no intentional God—at least not a very intelligent one. I also suffer some trepidation before religious people in general whenever discussing anything of moral substance, since it's long been my opinion that God is the Great Obfuscator, unnecessarily complicating many otherwise straight-forward humanistic matters.

So now that I've come out of the atheistic closet, entirely undressed, how can I possibly say that I trust those who believe in God more than those whom I'd otherwise con-sider to be sympathetic and like-minded thinkers? Well, trustworthiness is a different thing altogether from intellect, and I suppose I'm ever the social pragmatist in my dealings with other people.

Take, for example, a situation I found myself in outside a rail station in an Irish seaside town years ago. My luggage in hand, the cold gray sky windy and threatening to rain, I was confronted with two taxis at the curb waiting for passengers. One of the cars had a crucifix dangling from the rearview mirror and a dog-eared copy of the Bible on prominent display on the console. The other taxi showed no trace of any religious icons. Now, all else being equal, which of these two taxis would you choose, considering also that you're trying to avoid being overcharged, a practice for which this part of the country is notorious—and that being an American during the "W." administration, I might add, elevates you one step above our forty-third president in respectability? Both drivers are in all probability devout Catholics—this is Ireland, after all. Still, there's no way to know for certain.

Unless you're trying to make a point about how "atheists are good people too," or you happen to despise the Catholic Church, it's really a no-brainer: go with God. Why is this so obvious? As the political scientist Dominic Johnson has argued, "If supernatural punishment is held as a belief, then this threat becomes a deterrent in reality, so the mechanism can work regardless of whether the threat is genuine or not." In other words, from a psychological perspective, the ontological question of God's actual existence is completely irrelevant; all that really matters in the above case is that the taxi driver is fully convinced that God doesn't like it when he cheats his passengers.

This theoretical supposition that believers behave better because they feel that God is watching them, and presumably communicates His displeasure about their sinful deeds in the shape of various misfortunes, is one of the most compelling scientific arguments for the sheer stickiness of religion in society today. God just won't go away, and much of the reason He won't, goes this purely mechanistic evolutionary logic, is that the cognitive illusion of a punitive God functions to stem the selfish behaviors of individuals and helps to sustain social harmony.

A number of studies have offered empirical support for this supernatural monitoring hypothesis. This is a term coined by Ara Norenzayan, who in multiple studies has found that when participants are implicitly primed with God-related words ("spirit," "divine," "sacred," and so on), they become both more "prosocial" and less anti-social. By contrast with nonreligious or neutral words, people who see such religious words, for example, donate more money to a charity after completing a word-scramble task in which they cobble the words together into some coherent sentence. Although he and his collaborator Azim Shariff favored the interpretation that participants behaved more altruistically in the religious condition because the religious words reminded them that God was watching and therefore judging them, Norenzayan had always been cautious not to conclude prematurely that this was caused simply by concerns about heavenly spying. It's also possible, of course, that these religious words simply activated related social concepts such as "benevolence" and "good deeds," priming altruistic decision making independent of worrying about God's fretful glares.

More recent work, however, has allowed Norenzayan to put those concerns to rest. Getting people to think about God—even unconsciously and even, interestingly enough, among nonbelievers—indeed triggers very specific reasoning about their being the targets of someone's visual attention. Norenzayan and Will Gervais found that this basic effect of religious words making people feel visually exposed panned out across a variety of experimental conditions. In one study, for instance, the investigators used the same implicit God-priming method as before, assigning either a religious or a nonreligious word-scrambling task to believers and atheists. The participants then completed something called the Situational Self-Awareness Scale, and, remarkably, regardless of their explicit belief or disbelief in God, all those who'd been exposed unconsciously to the religious words—but not to the neutral words—showed a spike in their public self-awareness. That is to say, they became significantly more cognizant and concerned about the transparency of their social behaviors from an audience's point of view.

Furthermore, "when people feel that their behavior is being monitored," reason Norenzayan and Gervais in a follow-up experiment, "they tend to cast themselves in a positive light." This led them to hypothesize that reminders about God would not simply increase self-awareness but also encourage socially desirable responses. Participants' responses to statements such as "I am sometimes irritated by people who ask favors of me" and "No matter who I'm talking to, I'm always a good listener" should reflect their beliefs about what God *wants* to hear, not the *truth* about these unrealistically positive social attributes. In this study, however, the only people who produced socially desirable responses to the implicit God primes were those who actually believed in God. This means that while nonbelievers might feel "exposed" in the wake of receiving implicit God primes, just like believers, this feeling doesn't influence how atheists attempt to portray themselves socially.

For believers, in fact, additional evidence shows that God-related cues not only infuence their desire to have others see them in a positive way but actually motivate them to do good deeds. Some of the best support for this is the so-called Sunday Effect, first identified by Deepak

Malhotra of the Harvard Business School. Malhotra's research has also revealed how it's the context of the situation—particularly the presence or absence of ostensibly holy cues—that flushes out any actual differences in altruism between believers and nonbelievers. "This approach helps to shift away from seeking a simple answer to the question of whether religious people are nicer," reasons Malhotra in *Judgment and Decision Making,* "and towards assessing when, if ever, religious people may be nicer." Malhotra hypothesized that religious individuals would be more responsive to appeals from charities than would nonreligious people, *but only on days when they had earlier attended church.*

To test this prediction, the author collaborated with an online auction house that agreed to systematically alternate its scripted language for encouraging continued bidding. For online participants who'd been randomly assigned to the "charity-focused" message, the prompt read as follows:

We hope that you will continue to support this charity by keeping the bidding alive. Every extra dollar you bid in the auction helps us to accomplish our very important mission.

By contrast, people who'd been picked to receive the "competitive" message saw this:

The competition is heating up! If you hope to win, you will have to bid again.
Are you up for the challenge?

Importantly, Malhotra also had an independent measure of the bidders' religiosity, including their church attendance habits, which he obtained six weeks after they'd made their bidding decisions in response to one of these two primes. "The effect size is compelling," he explained. "On Sundays, appeals to charity were 300% more effective on religious individuals compared to non-religious individuals." By contrast, there was absolutely no difference between the religious and the nonreligious bidders in the effectiveness of the charity appeals on any other day of the week. There's another interesting Sunday Effect finding too, this one uncovered by chance by the economist Benjamin Edelman, also of the Harvard Business School. In crunching the salacious online numbers, Edelman discovered that the U.S. population is significantly less likely to purchase online subscriptions to pornographic websites on Sundays than they are on any other day of the week.

Although much of this may make for common sense, the fact that salient religious cues prompt neighborly decisions and curb social transgressions because they focus the believers' attention on God's hawkeyed view of their behaviors is tremendously important for understanding the adaptive *function* of religion. And such effects play out all around us. In many courtrooms across the Western world, for instance, defendants and witnesses must place their hand on the Bible and volunteer to respond to the religious oath "Do you swear to tell the truth, the whole truth, and nothing but the truth, so help you God?" And in the ancient Hebrew world, there was the similar "oath by the thigh"—where "thigh" was the polite term for one's dangling bits—since touching the sex organs before giving testimony was said to invoke one's family spirits (who had a vested interest in the seeds sprung from these particular loins) and ensured that the witness wouldn't perjure him-self. I rather like this older ritual, in fact, as it's more in keeping with evolutionary biology. But in general, swearing to God, in whatever way it's done, is usually effective in persuading others that you're telling the truth. We know from controlled studies with mock juries that if a person swears on—or, better yet, kisses—the Bible before testifying, the jury's perception of that person's believability is significantly enhanced.

After all, who in their right mind would lie before God? Well, as these findings suggest, atheists are more likely to do so. And that's the reason—the only reason—that I'd choose a Catholic taxi driver in Ireland over one who, like me, thinks that little book on the other driver's console is filled with nonsense of papal proportions.

God's Little Rabbits: Believers Outreproduce Nonbelievers by a Landslide

What's that famous quotation by Edna St. Vincent Millay? Oh, yes. I remember now: "I love Humanity; but I hate people." It's a sentiment that captures my normal misanthropically tinged type of humanitarianism well, but it becomes roaringly apropos on some particular occasions. For example, while I was making conversation at the pizza shop in a small village in Northern Ireland, the topic turned to what I did for a living. Now, this simple query was usually a hard question for me to answer; when I said I'm a professor, inevitably I was asked what I taught. When I said psychology, people giggled uncomfortably about their problems or replied—as if it were the most original line—that I'm in the right town for that. When I corrected them and said I'm not a clinical psychologist but a researcher, I had to explain what exactly I research.

"Evolutionary psychology" tended to conjure up some bizarre ideas in many people's minds. And so it did on this occasion, as I struggled to articulate the nature of my profession in a cramped pizza parlor with about half a dozen locals eavesdropping as I did so. Somehow or another, as conversations with me so often do, homosexuality came up as an example of a complex human behavior that evolutionary psychologists are still trying to understand.

I wish I'd had a notebook in hand to scribble down the young employee's comments word for word, so as to provide you with a proper ethnographic account. But here, in a nutshell, is what he very confidently said to me, flavored with the peculiar vernacular flourish found in this part of the world: "Aye. Don't get me wrong, I've got nothin' against gay people. But what I don't get is why they'd choose to be selfish and not 'ave a family and kids—like which is what we're here for, how's you's go against evolution by not continuin' the line 'cause you's can't help the species without having kids. Just seems selfish-like to me." I replied that as a gay man myself, it's not quite as simple as "choosing" not to reproduce; since women are about as arousing to me as that half-eaten pepperoni pizza sitting on that table over there, I said, I couldn't get an erection to inseminate a woman for the life of me. I do, however, I continued, get a mighty erection by seeing other men's erections, so therein—I pointed my finger to the heavens for emphasis—lies the true Darwinian mystery! I then took my pizza and left. In haste. And now I'm writing this from Ohio.

But in any event, the exchange reminded me of the German sociologist Michael Blume's research on reproduction and religiosity. And it occurred to me that religiously motivated homophobia may be at least partially rooted in this assumption that gay people are shirking their human reproductive obligations. I detected a strong whiff of religious residue in the employee's comments about homosexuality, which given the churchliness of Northern Ireland probably wasn't my imagination.

In evolutionary biological terms, where natural selection occurs at the level of the gene, not at the level of the species, there are serious flaws in this person's conjecture about lineal reproduction. Modern technological methods helping gays to be parents aside, there are many ways that childless individuals can still be genetically successful, in some cases more so than simply being a biological parent, such as investing heavily in biological kin who share their genes. (In scientific parlance, this is known as kin selection or inclusive genetic fitness.) Having said that, I'll acknowledge he was not entirely wrong about the prime evolutionary significance of reproduction either. People really do need to reproduce, either directly or indirectly, for nature to continue operating on their genes. This is not the "reason" or "purpose" we're here, as that would insinuate some form of intelligent design for human existence; rather, it's just a mechanical fact.

But all of this gets really interesting, says Blume, where the illusion of intelligent design intersects with a reproductive imperative—essentially, the commonplace idea that God "wants" or "intends" or "demands" us, as faithful members of our communities, to have a litter of similarly believing children. You've been blessed with your pleasure-giving loins for a reason, so the logic goes, and that's to get married to the opposite sex and to breed. By God, just look at the Old Testament. "Be fruitful and multiply" is the very first of 661 direct commandments. God seems to be not merely making a suggestion here but issuing a no-nonsense order.

Blume has found that those religions that actually put this issue front and center in their teachings are—for rather obvious reasons—at a selective group advantage over those that fail to endorse this stern commandment. He reviews several religions that are either already extinct or presently disappearing because they strayed too far from this reproductive principle. The Shakers, for example, hindered and even forbade reproduction among their own followers, instead placing their emphasis on missionary work, proselytizing, and the conversion of outsiders. But this turned out to be a foolish strategy, evolutionarily speaking. "In the long run," Blume points out, "mass conversions happen to be the historic exception, not the rule. Most of the time, only fractions of populations tend to convert from the religious mythology handed to them vertically by their parents and they convert into different directions ... Communities who start to lack young members also tend to lose their missionary appeal to other young people. Therefore, the Shakers overaged and deteriorated."

Some religious splinter groups have also tinkered a bit too much with God's reproductive imperative, even exploring eugenics by attempting to "perfect" communal offspring. Such a calculated, deliberate scheme of human breeding may backfire, however, if it also means preventing couples from reproducing at their own personal discretion. This was part of the downfall of the Oneida Community of upstate New York, a nineteenth-century Christian commune that had a very practical—almost too practical—view of human sexuality. Reproduction was tightly regulated by a eugenics system known as stirpiculture. Over several generations, Oneida Community physicians mated men and women who were carefully selected for their genetic health (I saw some of the handwritten medical records while going through the archives at the Kinsey Institute, and I can assure you that the breeding system was real and meticulous). Children born through this process of artificial selection were raised communally, and maternal bonding was discouraged.

To prevent unplanned, nonengineered children, the Oneida members implemented a variety of controls, including encouraging teenage boys to have sex with post-menopausal women. This simultaneously stemmed both parties' libidos and, in forging personal alliances between the two, provided important ecumenical tutelage to the youth by the very devout older women. Adult men practiced male continence, a sexual "technique" in which males do not ejaculate during intercourse; given that Oneida also had polyamorous relationships, this was key for stirpiculture purposes.

All of this may sound logical in theory, even unusually rational as far as religions go, but the tight regulations meant a quick death for the Oneida Community. After only about thirty years and peaking at just a couple of hundred members, the religious commune officially dissolved in 1881. Its members, presumably of good genetic stock but scanty in ranks, went into the silverware trade instead; today the Oneida Community is known as the hugely successful company Oneida Limited.

By contrast, similarly insulated, nonproselytizing religions that encourage their members to proliferate alleles the old-fashioned way—such as Orthodox Jews, the Hutterites, and the Amish—and also emphasize "home-grown" faith in which members are born into the group and indoctrinated, are thriving. The story of the Amish is particularly impressive, having seen an exponential explosion in their numbers over a very short span of time. The Amish emerged as a branch of the Anabaptist movement in the aftermath of the Protestant Reformation in Europe, and about four thousand of them fled Germany to avoid persecution and found refuge in the United States and Canada during the eighteenth and early nineteenth centuries. Most people know that the Amish are extremely insular, shunning almost all contact with the non-Amish world—except during the brief *Rumspringa* (or "jumping around") period, in which not-yet-baptized Amish youth flirt with the devilish goods outside before deciding whether or not to return to their family and faith. For boys, one incentive for returning to the community is that if you want to have sex with (that is, marry) a local Amish girl, you have to be baptized first, which is only for those who come home. Eighty percent do.

What you may not know is that the Amish population has been swelling since the sect's arrival in the New World. With growth rates hovering between 4 and 6 percent per year, their numbers double every twenty years or so. In 2008 they numbered 231,000; the year before, it was 218,000. Having children is a heavenly blessing, but it's also an official duty. With an average of six to eight children born to each Amish woman, and with 80 percent of those returning to the group after their *Rumspringa*, this extraordinary growth rate is easy to understand. What's especially ironic, Blume points out, is that the Amish's original country of origin, Germany, has been succumbing to sharp population declines for decades: "The closing of churches has been followed by that of playgrounds, kindergartens, schools and whole settlements." At least in sheer numbers, then, it seems that the Amish—long ridiculed by their European countrymen as the "dumb Germans" who wouldn't give up their silly archaic beliefs—are having the last laugh.

In fact, Blume's research also shows quite vividly that secular, nonreligious people are being dramatically outreproduced by religious people of any faith. Across a broad swath of demographic data relating to religiosity, the godly are gaining traction in offspring produced. For example, there's a global-level positive correlation between frequency of parental worship attendance and number of offspring. Those who "never" attend religious services bear, on a worldwide average, 1.67 children per lifetime; "once per month," and the average goes up to 2.01 children; "more than once a week," 2.5 children. Those numbers add up—and quickly.

Some of the strongest data from Blume's analyses, however, come from a Swiss Statistical Office poll conducted in the year 2000. These data are especially valuable because nearly the entire Swiss population answered this questionnaire—6,972,244 individuals, amounting to 95.67 percent of the population—which included a question about religious denomination. "The results are highly significant," writes Blume: "Women among all denominational categories give birth to far more children than the non-affiliated. And this remains true even among those (Jewish and Christian) communities who combine nearly double as much births with higher percentages of academics and higher income classes as their non-affiliated Swiss contemporaries."

In other words, it's not just that "educated" or "upperclass" people have fewer children and tend also to be less religious, but even when you control for such things statistically, religiosity independently predicts the number of offspring born to mothers. Even flailing religious denominations that place their emphasis on converting outsiders, such as Jehovah's Witnesses, are outreproducing nonreligious mothers. Hindus (2.79 births per woman), Muslims (2.44), and Jews (2.06), meanwhile, are prolific producers of human beings. Nonreligious Swiss mothers bear a measly 1.11 children.

Blume recognizes, of course, the limits in inferring too much from these data. It's not entirely clear whether being religious causes people to have more children, or whether—as is somewhat less plausible but also possible—the link is being driven in the opposite direction (with people who have more children becoming more religious). Most likely, it's both. Nevertheless, Blume speculates on some intriguing causal pathways tied to the fact that religious people have more children. We know from twin studies, for example, that the emotional components of religiosity are heritable. "Religiosity" refers to the *intensity* of feelings about religion, not the propositional content of particular beliefs. (In other words, one identical twin might be a screaming atheist, while the other is an evangelical pastor, but they're both hot and bothered by God.) So Blume surmises that any offspring born to religious parents are not only dyed in the wool of their faith through their culture but also genetically more susceptible to indoctrination than are children born to nonreligious parents.

The whole situation doesn't bode well for secularist movements, in any event. Evolutionary biology works by a law of numbers, not rational sentiments. Blume, who doesn't try to hide his own religious beliefs, sees the cruel irony in this as well: "Some naturalists are trying to get rid of our evolved abilities of religiosity by quoting biology. But from an evolutionary as well as philosophic perspective, it may seem rather odd to try to defeat nature with naturalistic arguments."

As a childless gay atheistic soul born to a limply interfaith couple, I suspect, perhaps for the better, that my own genes have a very mortal future ahead. As for the rest of you godless heterosexual couples reading this, toss your contraceptives and get busy in the bedroom. Either that or, perish the thought, God isn't going away any time soon.

Planting Roots with My Dead Mother

Mother's Day is forever tinged with a certain sadness for me because it's the day I accompanied my mother many years ago to the cemetery where she's been interred ever since. Well, that's not entirely true. She didn't die that very day; death wouldn't come for another six months yet.

We were in the funeral home to shop for a shiny new casket and make final arrangements for her corpse, an unwelcome visitor that would be arriving some time soon, though precisely when even the doctors couldn't say. For her peace of mind if nothing else, she was intent on tidying up the financial and administrative minutiae that come with dying as a human being. As soon as the umbilical cord is cut, after all, we're attached to another made of red tape, and that one grows longer with each passing year, so that we die tangled up in it in the end.

I don't know why she chose Mother's Day of all days for such a lachrymal task as this, but she did have a tragedian's air to her—one, I might add, that was well deserved given all she'd been through. Before she was forty, she'd had a mastectomy from breast cancer along with several long bouts of chemotherapy. This was followed by cancer in the other breast a few years later and another mastectomy. Within the decade, my parents would have a sudden and bitter divorce, and within a few months of the divorce, just as she was "getting back on her feet," she was dealt another heavy blow, diagnosed with late-stage ovarian cancer, and had to undergo more surgeries and seven more embattled years of chemotherapy. She died—begrudgingly—at just fifty-four.

It's a very sad story, needless to say, and unfortunately one that is shared by too many other loving and wonderful mothers who will not be with us on the next Mother's Day. The fact that I was conducting research on people's afterlife beliefs at the time of her death stemmed almost entirely from the many theoretically inspiring and insightful conversations I had with her as she tried to imagine her own afterlife. (She leaned toward scientific materialism, but she wasn't an atheist and had an open mind about the whole affair, I think it's safe to say.)

Among the more unpleasant aspects of this tale—both for her at the time and for my siblings and me still now—were the gloomy logistics of arranging her burial. What sticks out in my mind most of all from that Mother's Day 2000 is the image of my mom with her trembling hands flipping through an L.L.Bean-type catalog handed to her by a pleasant enough but benumbed funeral home director. It was a rather hefty booklet filled with glossy images of all the latest models of caskets, vaults, urns, catafalques, headstones, and other new products then in funerary vogue, this particular collection especially suitable for middle-class cadavers. Since she died near Fort Lauderdale wanting to be closer to her own mother, she found herself in a part of the country especially profitable to the death industry, the area being a geographic hub of the elderly.

The whole affair that day left a bad taste in my mouth. There was something so plastic, so slick, so "commercial" about this business of death that—much like the rest of an overdeveloped South Florida where this bland, freeway-hugging cemetery is laid—felt much too cold to me. Modern cemeteries, with their zero lot lines, perfectly manicured hedgerows, and identical-looking headstones, have become eerily similar to the suburbs; or perhaps the suburbs have become eerily similar to cemeteries. Either way, what bothers me most of all is that, looking back, it didn't have to be like this.

Death is rarely pleasant, of course, no matter how one's body is disposed of. But in recent years, I have become increasingly interested in "green burial," a blanket term that refers to any "alternative" funerary practice in which the deceased is buried in a biodegradable casket or shroud, often in nature preserves, and without embalming preservatives (fluids that keep a corpse pretty, usually just for viewing purposes) that dramatically slow down and disrupt the natural decomposition process.

Although it's the subject of continuing debate and the actual health implications remain unclear, these embalming chemicals may become contaminants as formaldehyde and other potentially carcinogenic agents are absorbed into the soil and groundwater. Green burial advocates have cast the issue almost entirely in terms of avoiding the staggering environmental impact of traditional burial. Consider that before this year is over, Americans will bury 827,060 gallons of embalming fluid, 90,272 tons of steel (caskets), 2,700 tons of copper and bronze (caskets), 1,636,000 tons of reinforced concrete (vaults), 14,000 tons of steel (vaults), and more than 30 million board feet of hardwoods (much of it tropical; caskets). Then there are the countless acres of land bulldozed over for these bald landfills of synthetic human remains.

Cremation isn't much of an improvement over such things. Going up in smoke may use fewer natural resources than traditional burial, but it also consumes a significant amount of fossil fuels. According to a statement by the Trust for Natural Legacies, a nonprofit land conservation organization working to drive the sustainable growth of green burial practices in the Midwest, "You could drive about 4,800 miles on the energy equivalent of the energy used to cremate someone—and to the moon and back 83 times on the energy from all cremations in one year in the U.S." There's also the non-negligible problem of mercury being released into the atmosphere whenever a person with amalgam dental fillings is cremated.

These environmental concerns alone make green burial an obvious solution. But as a psychologist, and one who's also had a negative personal experience with burying a loved one in the traditional manner, I think our conception of death and burial needs a serious rethink. Let's close the lid on those anonymous, revenue-driven, laminated cultural practices of commercial burial that we've all become so complacent about. There's got to be a better way to go about it than what we've been doing all these years. And one specific form of green burial, which I'll outline in a moment, is win-win.

Although the idea of green burials in wildlife preserves or parklike settings is not new, and it's likely a desirable prospect for certain future dead souls who'd prefer absolute oblivion, it seems to me that this is not going to appeal to most individuals because we human beings tend to have a pressing need for "symbolic immortality." This was a term coined by the cultural anthropologist Ernest Becker in his book *The Denial of Death*, but it has since been empirically elaborated by scientists working on terror management theory. The basic idea behind symbolic immortality is that cultural artifacts that survive the individual's literal death while also containing some reminder of the person's special existence can meaningfully reduce human death anxiety.

There are many nuances to terror management theory and this construct, but the important point to mention here is that a sense of symbolic immortality can be obtained by concrete markers of prosperity, anything from benches in the park with dead people's names etched in gold, to graffiti on boxcars, to initials carved into a tree, to headstones in a graveyard. So while conventional cemeteries may be unnecessarily gloomy, they do at least satisfy this psychological need for people to remain embedded, even if just symbolically by way of lifeless granite headstones, in the immortal culture. If the green burial industry is ever to take off and begin appealing to more people, I suspect that this is one key issue—physical memorializing—that advocates are going to need to address.

It seems to me that one way to solve this problem while remaining true to the central philosophy of green burial is to have people buried beneath a specific tree—a little sapling of your choice nourished by your decomposing body beneath. In favorable soil conditions, a non-embalmed body, skeleton and all, can rot away entirely within about fifteen to twenty-five years. But many trees species, let's not forget, can live for many hundreds of years (some thousands). Imagine that on making final arrangements at the funeral home someday, you and your loved ones were able to choose from among a wide variety of co-habitable tree species to find *just the right tree* to suit your fabulously unforgettable being—this instead of flipping through a catalog filled with caskets, coffins, and crypts as my mother found herself doing. Not only will your death nourish a new life, but you're also saving another tree, the one that would be sacrificed for your sake in the shape of a mass-produced coffin with plastic handles.

In addition to offering a healthy dose of symbolic

immortality, this form of specific-tree burial would tap into another central aspect of our psychology. In recent years, researchers have found that human beings operate with a strong essentialist bias. We tend to reason implicitly, and often explicitly, as though a person's unobservable "essence" were transmitted through physical contact with that individual. You'd probably cringe to think of wearing a child molester's eyeglasses, or a serial murderer's laundered T-shirt, but have trouble articulating precisely why donning such material causes you so much aversion. Likewise, you may have your deceased grandmother's wedding ring or the old jersey of your favorite football player stashed away somewhere, and these objects are cherished because they're so intimately linked to these adored individuals. In the present context, let's say that you buried your beloved dog beneath a rosebush in your garden. If you're anything like me, you'd have a special affinity for that particular rosebush over others, and it would be especially unpleasant should, say, someone uproot it and dangle it before you.

Now picture an entirely new brand of cemetery, a planned, verdant, protected land tended by trained arborists and filled not with row after row of bland, lifeless, crumbling headstones but instead with row after row of living trees. Each tree, selected for regional appropriateness and other suitability factors as advised by arborist staff, would symbolize a unique human existence. (Not to get carried away, but perhaps a plaque or marker might be added too, enhancing the symbolic immortality element, but aesthetics would of course vary.) These aren't simply trees planted in memoriam of the dead but leafy hybrids whose veins have absorbed individual human lives.

I'll go out on a limb here and say that even if one doesn't believe in some ethereal or religious version of the afterlife, it's rather difficult to escape the cognitive illusion that the unobservable essence of each person has been somehow gradually transmuted into his or her individual tree. Two massive walnut trees growing side by side with interlocking branches seem somehow more than mere trees when we learn that they're actually growing upon what was once a husband and wife who lived centuries before. There's no shortage of idyllic essentialist images like this—grandchildren climbing up their great-grandfather's limbs, children who'd been sickly in life now bursting with the blazing colors of autumn, beauty queens forever fragrant with immaculate cherry blossoms, still-born infants now magnificent oaks. It would take some time, of course, for this human arboretum to fully mature. But what's the rush?

In fact, our species' notorious difficulty in imagining our own psychological nonexistence is yet another cognitive factor that makes this particular form of green burial appealing. Since we have no proper analogy for the state-less state of death (we can't re-create consciously in our heads what it "felt like" to be under general anesthesia, or prior to our conception, or even during last night's dream-less, non-REM sleep), the closest we can get to mentally grasping what it will be "like" to be dead inevitably reifles nothingness.

With specific-tree burial, this simulation constraint principle of the afterlife finds a nonreligious, or even religious, outlet. For example, you might not believe that you've been literally reincarnated or reborn into the tree, but in envisioning its growth and rejuvenation year after year through all the socially active centuries of human affairs lying ahead, you'll find it rather difficult to refrain from attributing some of your own emotions to this living character of the tree.

It would sure be nice to hug a young palm tree in Florida this weekend. Of course, I'd have to worry about Mom's health anew, about her getting a nasty weevil infestation or perhaps being rudely split in two by a thunderbolt. But we'd have worked those "acts of God" into the contract, the funeral director and I.

PART VIII

Into the Deep: Existential Lab Work

Being Suicidal: Is Killing Yourself Adaptive? That Depends: Suicide for Your Genes' Sake (Part I)

Most psychological science is the science of being and feeling like a human being, and since there is only one human being that I have or ever will have experience in being, it is not always clear to me where my career ends and my personal life begins. And this was especially true for me recently because, like many other adult gay commentators and horrified onlookers, the raft of gay teen suicides in recent years has reawakened memories of my own adolescent battles with suicidal thoughts. There is so much I want to say about this, in fact, because I'm reminded of the many illuminating theories and studies on suicide I've come across that helped me to understand—and, more important, to overcome and to escape from—that frighteningly intoxicating desire to prematurely rid myself of a seemingly interminable hell.

If only I could have reached out and gotten hold of Rutgers University's Tyler Clementi's shirttail before he leaped off the George Washington Bridge, or eased my fingertips between the rope and the neck of thirteen-year-old Seth Walsh before he hanged himself from a tree in his backyard, I would have pointed out to them that one day, they will find beauty even in this fleeting despair. I would tell them that their sexual orientation places them in the company of some of the greatest figures and secular angels in creative history—to name just a few, Michelangelo, Caravaggio, Oscar Wilde, Andy Warhol, Leonardo da Vinci, Marcel Proust, Jean Genet, Hans Christian Andersen, and Tchaikovsky. Finally, I'd tell them about the scientific research and ideas that I'm going to share with you, razor-sharp reasoning by bright scholars that might have pierced their suicidal cognition just enough to allow them to breathe a little more easily through those suffocating negative emotions.

A scientific understanding of suicide is useful not only for vulnerable gay teens but for all those ever finding themselves in conditions favoring suicide. I say "favoring suicide" because there is convincing work—all tracing back to Denys deCatanzaro's largely forgotten ideas from the early 1980s—indicating that human suicide is an adaptive behavioral strategy that becomes increasingly likely to occur whenever there is a perfect storm of social, ecological, developmental, and biological variables factoring into the evolutionary equation. In short, deCatanzaro posited that human brains are designed by natural selection in such a way as to encourage us to end our own lives when facing certain conditions, because this was best for our suicidal ancestors' overall genetic interests.

For good-hearted humanitarians, it may sound rather bizarre, perhaps even borderline insensitive, to hear that suicide is "adaptive." But remember that this word means a very different thing in evolutionary terms than it does when used in clinical settings. Because natural selection operates only on phenotypes, not human values, even the darkest of human emotions may be adaptive if they motivate gene-enhancing behavioral decisions. It's not that evolution is cruel, but as a mindless mechanism it can neither care nor not care about particular individuals; selection, after all, is not driven by an actual brain harboring any feelings about, well, anything at all. In no case does this sobering fact come into sharper focus than with adaptive suicide.

Saying that suicide is adaptive may also sound odd to you from an evolutionary perspective, because on the surface it seems to fly in the face of evolution's first rule of thumb, which is to survive and reproduce. However, as the evolutionary theorist William Hamilton's famous principle of inclusive fitness elucidated so clearly, it is the proportion of one's genetic material surviving in subsequent generations that matters; and so if the self's survival comes at the expense of one's genetic kin being able to pass on their genes, then sacrificing one's life for a net genetic gain may have been adaptive ancestrally.

Before we get ahead of ourselves, let's first ease into the suicide-as-adaptation argument with a few nonhuman examples, which come mostly from the insect and arthropod worlds. Take male Australian redback spiders (*Latrodectus hasselti*), for instance, which seem content to be cannibalized by—to say the least—sexually aggressive female redback spiders during sex. Aside from putting a damper on an otherwise enjoyable act, being eaten alive while copulating would seem rather counterintuitive from an evolutionary perspective. But when biologists looked more closely at this spidery sex, they noticed that males that are cannibalized copulate longer and fertilize more eggs than males that are not cannibalized; and the more cannibalistic a female redback spider is, it turns out, the more desirable she is to males, even rejecting more male suitors than her less cannibalistic counterparts.

Another example is bumblebees (*Bombus lucorum*), a species that is often parasitized by invidious little conopid flies that insert their larvae in the bee's abdomen. Once infected, the bumblebee dies in about twelve days, and the parasitical flies pupate until their emergence the following summer. What's interesting about this, however, is that parasitized bumblebees essentially go off to commit suicide by abandoning their colony and spending their remaining days alone in faraway flowery meadows. In doing so, these infected bumblebees are leading the flies away from nonparasitized kin, increasing inclusive fitness by protecting the colony from infestation.

What is critical to take away from these nonhuman examples is that the suicidal organism is not consciously weighing the costs of its own survival against inclusive fitness gains. Redback spiders and bumblebees aren't mindfully crunching the numbers, engaging in self-sacrificial acts of heroic altruism, or waxing philosophical on their own mortality. Instead, they are just puppets on the invisible string of evolved behavioral algorithms, with neural systems responding to specific triggers. And, says deCatanzaro, so are suicidal human beings whose emotions sometimes get the better of them.

So let's turn our attention now to human suicide. To crystallize his position, I present deCatanzaro's "mathematical model of self-preservation and self-destruction" (circa 1986): $\Psi_i = \rho_i + \Sigma b_k \rho_k r_k$,

where Ψ_i = the optimal degree of self-preservation expressed by individual i (the residual capacity to promote inclusive fitness); ρ_i = the remaining reproductive potential of i; ρ_k = the remaining reproductive potential of each kinship member k; b_k = a coefficient of benefit (positive values of b_k) or cost (negative values of b_k) to the reproduction of each k provided by the continued existence of i ($-1 \leq b \leq 1$); r_k = the coefficient of genetic relatedness of each k to i (sibling, parent, child = .5; grandparent, grandchild, nephew or niece, aunt or uncle = .25; first cousin = .125; etc.).

For the mathematically disinclined, this can all be translated as follows: people are most likely to commit suicide when their direct reproductive prospects are discouraging and, simultaneously, their continued existence is perceived, whether correctly or incorrectly, as reducing inclusive fitness by interfering with their genetic kin's reproduction. Importantly, deCatanzaro, as well as other independent researchers, has presented data that support this adaptive model.

In a 1995 study in *Ethology and Sociobiology*, for example, deCatanzaro administered a sixty-five-item survey including questions about demographics (such as age, sex, and education), number and degree of dependency of children, grandchildren, siblings, and siblings' children, "perceived burdensomeness" to family, perceived significance of contributions to family and society, frequency of sexual activity, stability/intimacy/success of relations to the opposite sex, homosexuality, number of friends, loneliness, treatment by others, financial welfare and physical health, feelings of contentment, depression, and looking forward to the future. Respondents were also asked about their suicidal thoughts and behaviors—for example, whether they had ever considered suicide, whether they had ever attempted it in the past, or whether they ever intended to do so in the future. The survey was administered to a random sample of the general Ontario public but also to targeted groups, including elderly people from senior citizen housing centers, psychiatric inpatients from a mental hospital, male inmates incarcerated indefinitely for antisocial crimes, and, finally, exclusively gay men and women.

Many fascinating—and rather sad—findings emerged from this study. For instance, the greatest levels of recent suicide ideation were in male homosexuals and the psychiatric patients, whereas the prison population showed the most previous suicide attempts. "It gets better," sure, but we're always at risk, and this evolutionarily informed model helps gay individuals to face and understand that lamentable reality. But the important takeaway message is that the pattern of correlational data conformed to those predicted by deCatanzaro's evolutionary model. Although the author offers the important disclaimer that "the observational nature of this study limits strong causative inferences," nevertheless: "The profile of correlations agrees with the notion that suicidal ideation is related to conjunction of poor reproductive prospects and diminished sense of worth to family. Concordance of the data with the hypothesis is apparent in reliable relationships of reproductive and productive parameters to suicidal ideation."

One noteworthy thing to point out in such data is the meaningful developmental shift that occurs in the motivational algorithm. Whereas heterosexual activity is the best inverse predictor of suicidal thoughts among younger samples, this is largely replaced among the elderly by concerns about finances, health, and especially the sense of "perceived burdensomeness" to family. A few years after this *Ethology and Sociobiology* report, a followup study in *Suicide and Life-Threatening Behavior*, conducted by an independent group of investigators seeking to further test deCatanzaro's model, replicated the same predicted trends.

As persuasive as I find this model, I still had a question

left unanswered by deCatanzaro's basic argument, so I asked him for clarification. Basically, I wanted to know how the suicidal patterns of contemporary human beings relate to those of our ancestral relatives, who presumably faced the conditions in which the adaptation originally evolved, but who in many ways lived in a very different world from our own. After all, even with guns, knives, and drugs at our disposal, committing suicide is not always an easy thing to do, practically speaking.

In an article published in *Psychological Review*, for instance, the psychiatrist Kimberly Van Orden and her colleagues cite the case of a particularly tenacious suicidal woman: "[She] was described as being socially isolated when she attempted suicide with an unknown quantity and type of pain medication and also opened her wrist arteries. This action led to some degree of unconsciousness, from which she awoke . . . She then threw herself in front of a train, which was the ultimate cause of her death."

Now consider the suicide methods that would have been available to our ancient relatives in a technologically sparse environment—perhaps a leap from a great height that, if one weren't successful, might have at least led to wounds sufficient enough for the person to eventually die from infection. Starvation. Exposure. Drowning. Hanging. Offering oneself to a hungry predator. Okay, so maybe there were more methods available to our ancient forebears than I realized. You see what I mean, though. Today, moving your fingertip but a hairsbreadth on a trigger is a surer route to oblivion than anything our species has ever known before; gun owners might as well have an "off" button, it's so simple now. (This is one of the many reasons that I don't own a gun; deCatanzaro's suicide algorithm is stochastic, which means that the figure it generates for a given individual is in a constant state of flux.)

But deCatanzaro doesn't see technological advances as particularly problematic for his model. Fossils of suicidal australopithecines or early *Homo sapiens* aren't easy to come by, of course. But as he wrote to me:

> Evidence indicates appreciable rates of suicide throughout recorded history and in almost every culture that has been carefully studied. Suicide was apparently quite common in Greek and Roman civilizations. Anthropological studies indicate many cases in technologically primitive cultures as diverse as Amerindians, Inuit, Africans, Polynesians, Indonesians, and less developed tribes of India. Self-hanging was one of the most prevalent methods of suicide in such cultures. There are also data from developed countries comparing suicide rates from the late nineteenth century through the twentieth century. These data show remarkable consistency in national suicide rates over time, despite many technological changes. So, the data actually do not show a major increase in suicide in modern times, although this inference must be qualified in that there may have been shifts in biases in recording of cases. Interestingly, the methods of suicide have changed much more than the rates. For example in Japan, hanging prevailed until 1950, after which pills and poisons became the primary method. In England and Wales, hanging and drowning were common in the late 19th century, but were progressively replaced by drugs and gassing. Motives may have been more constant than means.

I find deCatanzaro's argument that suicide is adaptive both convincing and intriguing. But I do think it begs for more follow-up research. For example, his inclusive fitness logic should apply to every single social species on earth, so why is there such an obvious gap between frequency of suicide in human beings and other animals? Each year, up to twenty million people worldwide attempt to commit suicide, with about a million of these completing the act. That's a significant minority of deaths—and near deaths—in our species. And there is reason to be suspicious that nonhuman animal models (such as parasitized bumble-bees, beached whales, leaping lemmings, and grieving chimpanzees) are good analogues to human suicide. In our own species, suicide usually means deliberately trying to end our psychological existence—or at least this particular psychological existence. And whereas most other accounts of "self-destruction" in the natural world seem to involve some type of interspecies predation or parasitical manipulation, human suicides are more often driven by negative interpersonal appraisals made by other members of our own species. In fact, Robert Poulin, the zoologist who first reported on the altered behavior of those parasitized bumble-bees, even urges researchers to use caution in referring to such examples as "suicide": "The adoption of a more dangerous lifestyle by an insect that is bound to die shortly may be adaptive in terms of inclusive fitness, but no more suicidal than, for instance, an ageing animal taking risks to reproduce in the presence of a predator as its inevitable death approaches."

I believe that suicide, like fantasy-enabled masturbation, requires evolved social cognitive processes that are relatively unique—in this case, painfully so—to our species. There are anecdotes aplenty, of course, but there are no confirmed cases of suicide in any nonhuman primate species. Although there are certainly instances of self-injurious behaviors, such as excessive self-grooming, these are almost always limited to sad or abnormal social environments such as biomedical laboratories and zoos. Yes, grieving young chimps have been known to starve to death from depression in the wake of their mother's death, but there is no evidence of direct self-inflicted lethal displays in monkeys and apes. Perhaps Jane Goodall can correct me if I'm wrong about this, but as far as I'm aware, there are no cases in which a chimpanzee has been observed to climb the highest branch it could find—and jump.

I think part of the answer to this cross-species mystery can be found in another theoretical model of suicide, this one by the psychologist Roy Baumeister, which I've always viewed as the "proximate" level to deCatanzaro's "ultimate" level of explanation for suicide. These are not alternative accounts of human suicide but deeply complementary ones. While deCatanzaro explains suicide in terms of evolutionary dynamics, Baumeister zeros in on the specific psychological processes, the subjective lens through which a suicidal person sees the world. His model describes the engine that actively promotes the adaptive response of suicide. I should hasten to add that I don't think either of them—deCatanzaro or Baumeister—necessarily sees his model as being complementary to the other's in this way. I don't even know if either is aware of the other. But this is how the two approaches have always struck me. Baumeister's take on the subject is, quite honestly, one of the most shockingly insightful I have ever read, in any research literature. In part II of our look at suicide and psychology, we'll turn our attention to that work.

Being Suicidal: What It Feels Like to Want to Kill Yourself (Part II)

One of the more fascinating psychotic conditions in the medical literature is known as Cotard's syndrome, a rare disorder, usually recoverable, in which the primary symptom is a "delusion of negation." According to the researchers David Cohen and Angèle Consoli of the Université Pierre et Marie Curie, many patients with Cotard's syndrome are absolutely convinced, without even the slimmest of doubts, that they are already dead.

Some recent evidence suggests that Cotard's may occur as a neuropsychiatric side effect in patients taking the drug acyclovir or valacyclovir for herpes and who also have kidney failure. But its origins go back much further than these modern drugs. First described by the French neurologist Jules Cotard in the 1880s, the syndrome is usually accompanied by some other debilitating problem, such as major depression, schizophrenia, epilepsy, or general paralysis—not to mention disturbing visages in the mirror. Consider the case of one young woman described by Cohen and Consoli: "The delusion consisted of the patient's absolute conviction she was already dead and waiting to be buried, that she had no teeth or hair, and that her uterus was malformed."

Poor thing—that image couldn't have been very good for her self-esteem. Still, call me strange, but I happen to find a certain appeal in the conviction that one is, though otherwise lucid, nevertheless already dead. Provided there were no uncomfortable symptoms of rigor mortis cramping up my hands, nor delusory devils biting at my feet, how liberating it would be to be able to write like a dead man and without that hobbling, hesitating fear of being unblinkingly honest. Knowing that upon publication I would be tucked safely away in my tomb, I could finally say what's on my mind. Of course, living one's life as though it were a suicide note incarnate (yet remember this is precisely what life is, really, and I would advise any thinking person to stroll by a cemetery each day, gaze unto those fields of crumbling headstones filled with chirping crickets, and ponder, illogically so, what these people wish they might have said to the world when it was still possible for them to have done so) is an altogether different thing from the crushing, unbearable weight of an actual suicidal mind dangerously tempted by the promise of permanent quiescence.

In considering people's motivations for killing themselves, we need to recognize that most suicides are driven by a flash flood of strong emotions, not rational, philosophical thoughts in which the pros and cons are evaluated critically. And, as I mentioned in the previous chapter on the evolutionary biology of suicide, from a psychological-science perspective, I don't think any scholar ever captured the suicidal mind better than the psychologist Roy Baumeister in his 1990 *Psychological Review* article, "Suicide as Escape from Self." To reiterate, I see Baumeister's cognitive rubric as the engine of emotions driving deCatanzaro's biologically adaptive suicidal decision making. There are certainly more recent theoretical models of suicide than Baumeister's, but none in my opinion are an improvement. The author gives us a uniquely detailed glimpse into the intolerable and relentlessly egocentric tunnel vision that is experienced by a genuinely suicidal person.

According to Baumeister, there are six primary steps in the escape theory, culminating in a probable suicide when all criteria are met. I do hope that having knowledge about the what-it-feels-like phenomenology of "being" suicidal helps people to recognize their own possible symptoms of suicidal ideation and—if indeed this is what's happening—enables them to somehow derail themselves before it's too late. Note that it is not at all apparent that those at risk of suicide are always aware that they are in fact suicidal, at least in the earliest cognitive manifestations of suicidal ideation. And if such thinking proceeds unimpeded, then keeping a suicidal person from completing the act may be as futile as encouraging someone at the very peak of sexual excitement to please kindly refrain from having an orgasm, which is itself sometimes referred to as *la petite mort* (the little death).

So let's take a journey inside the suicidal mind, at least as it's seen by Roy Baumeister. You might even come to discover that you've actually set foot in this dark psychological space before, perhaps without knowing it at the time.

Step 1: Falling Short of Standards. Most people who kill themselves actually lived better-than-average lives. Suicide rates are higher in nations with higher standards of living than in less prosperous nations; higher in U.S. states with a better quality of life; higher in societies that endorse individual freedoms; higher in areas with better weather; in areas with seasonal change, they are higher during the warmer seasons; and they're higher among college students who have better grades—and parents with higher expectations.

Baumeister argues that such idealistic conditions actually heighten suicide risk because they often create unreasonable standards for personal happiness, thereby rendering people more emotionally fragile in response to unexpected setbacks. So, when things get a bit messy, such people, many who appear to have led mostly privileged lives, have a harder time coping with failures. "A large body of evidence," writes the author, "is consistent with the view that suicide is preceded by events that fall short of high standards and expectations, whether produced by past achievements, chronically favorable circumstances, or external demands." For example, simply being poor isn't a risk factor for suicide. But going rather suddenly from relative prosperity to poverty has been strongly linked to suicide. Likewise, being a lifelong single person isn't a risk factor either, but the transition from marriage to the single state places one at significant risk for suicide. Most suicides that occur in prison and mental hospital settings happen within the first month of confinement, during the initial period of adjustment to loss of freedom. Suicide rates are lowest on Fridays and highest on Mondays; they also drop just before the major holidays and then spike sharply immediately after the holidays. Baumeister interprets these patterns as consistent with the idea that people's high expectations for weekends and holidays materialize, after the fact, as bitter disappointments.

To summarize this first step in the escape theory, Baumeister tells us that "it is apparently the size of the discrepancy between standards and perceived reality that is crucial for initiating the suicidal process." It's the proverbial law of social gravity: the higher you are to start off with, the more painful it's going to be when you happen to fall flat on your face.

Step 2: Attributions to Self. It is not just the fall from grace alone that's going to send you on a suicidal tailspin. It's also necessary for you to loathe yourself for facing the trouble you find yourself in. Across cultures, "self-blame" or "condemnation of the self" has held constant as a common denominator in suicides. Baumeister's theory accommodates these data, yet his model emphasizes that the biggest risk factor isn't chronically low self-esteem per se, but rather a relatively recent demonization of the self in response to the negative turn of events occurring in the previous step. People who have low self-esteem are often misanthropes, he points out, in that while they are indeed self-critical, they are usually just as critical of other people. By contrast, suicidal individuals who engage in negative appraisals of the self seem to suffer the erroneous impression that other people are mostly good while they themselves are bad. Feelings of worthlessness, shame, guilt, inadequacy, exposure, humiliation, or rejection lead suicidal people to dislike themselves in a manner that, essentially, isolates them from from an idealized humanity. The self is seen as being enduringly undesirable; there is no hope for change, and the core self is perceived as being rotten.

This is why adolescents and adults of minority sexual orientations, who grow up gestating in a social womb filled with messages—both implicit and explicit—that they are essentially lesser human beings, are especially vulnerable to suicide. Even though we may consciously reject these personal attributions made by an intolerant society, they have still seeped in.

Step 3: High Self-Awareness. Most scholars paint the emergence of self-awareness as a milestone achievement for our species. But with it comes the crushing truth of how we, as individuals, stack up to others. "The essence of self-awareness is comparison of self with standards," writes Baumeister. And, according to his escape theory, it is this ceaseless and unforgiving comparison with a preferred self—perhaps an irrecoverable self from a happier past or a goal self that is now seen as impossible to achieve in light of recent events—fueling suicidal ideation.

These unforgiving and unremitting thoughts in suicidal individuals are actually measurable, at least indirectly, by analyzing the language used in suicide notes. One well-known "suicidologist," Edwin Shneidman, once wrote, "Our best route to understanding suicide is not through the study of the structure of the brain, nor the study of social statistics, nor the study of mental diseases, but directly through the study of human emotions described in plain English, in the words of the suicidal person." Personally, I feel a bit like an existential Peeping Tom in reading strangers' suicide notes, but it's a long-standing practice in psychological research. Over the past few decades alone, nearly three hundred studies on suicide notes have been published. These cover a broad range of research questions, but because they tend to yield inconsistent findings, they have also painted a confusing picture of the suicidal mind.

This is especially the case when trying to reveal people's motivations for the act. Some who commit suicide may not even be aware of their own motivations, or at least they may not have been completely honest in their farewell letters to the world. A good example comes from the sociologist Susanne Langer and her colleagues' report in a 2008 issue of *The Sociological Review*. The researchers describe how the suicide note written by one young man was rather nondescript, mentioning feelings of loneliness and emptiness as causing his suicide, while, in fact, "his file contained a memo inquiring about the state of an investigation regarding sexual offences the deceased had been accused of in an adjacent jurisdiction."

The more compelling studies on suicide notes, in my view, are those that use text-analysis programs enabling the investigators to make exact counts of particular kinds of words. Compared with pretend suicide notes written as an exercise and "as though" one were about to kill oneself, real suicide notes are notorious for containing first-person-singular pronouns, a reflection of high self-awareness. And unlike letters written by people facing involuntary death, such as those about to be executed, suicide note writers rarely use inclusive language such as the plural pronouns us and we. When they do mention significant others, suicide note writers usually speak of them as being cut off, distant, separate, not understanding, or opposed. Friends and family, even a loving mother at arm's length, feel endless oceans away.

Step 4: Negative Affect. It may seem to go without saying that suicides tend to be preceded by a period of negative emotions, but, again, in Baumeister's escape model, negative suicidal emotions are experienced as an acute state rather than a prolonged one. "Concluding simply that depression causes suicide and leaving it at that may be inadequate for several reasons," he writes. "It is abundantly clear that most depressed people do not attempt suicide and that not all suicide attempters are clinically depressed."

Anxiety—which can be experienced as guilt, self-blame, threat of social exclusion, ostracism, and worry—seems to be a common strand in the majority of suicides. We may very well be the only species for which negative social-evaluative appraisals can lead to shame-induced suicide. The most convincing data from studies with nonhuman animals suggest very strongly that we are the only species on the face of the earth able to take another organism's perspective in judging the self's attributes. This is owed to an evolutionary innovation known as theory of mind (literally, theorizing about what someone else is thinking about, including what that person is thinking about you; and, perhaps more important in this case, even what you're thinking about you) that has been both a blessing and a curse. It's a blessing because it allows us to experience pride, and it's a curse because it also engenders what I consider to be the uniquely human, uniquely painful emotion of shame. (You'll also remember this from our earlier discussion on psychodermatology and acne.)

Psychodynamic theorists often postulate that suicidal guilt seeks punishment, and thus suicide is a sort of self-execution. But Baumeister's theory largely rejects this interpretation; rather, in his model, the appeal of suicide is loss of consciousness, and thus the end of psychological pain being experienced. And since cognitive therapy isn't easily available—or seen as achievable—by most suicidal people, that leaves only three ways to escape this painful self-awareness: drugs, sleep, and death. And of these, only death, nature's great anesthesia, offers a permanent fix.

Step 5: Cognitive Deconstruction. The fifth step in the escape theory is perhaps the most intriguing, from a psychological perspective, because it illustrates just how distinct and scarily inaccessible the suicidal mind is from that of our everyday cognition. With cognitive deconstruction, a concept originally proposed by the social psychologists Robin Vallacher and Daniel Wegner, the outside world becomes a much simpler affair in our heads—but usually not in a good way.

Cognitive deconstruction is pretty much just what it sounds like. Things are cognitively broken down into increasingly lowlevel and basic elements. For example, the time perspective of suicidal people changes in a way that makes the present moment seem interminably long; this is because "suicidal people have an aversive or anxious awareness of the recent past (and possibly the future too), from which they seek to escape into a narrow, unemotional focus on the present moment." In one interesting study, for example, when compared with control groups, suicidal participants significantly overestimated the passage of experimentally controlled intervals of time by a large amount. Baumeister surmises, "Thus suicidal people resemble acutely bored people: The present seems endless and vaguely unpleasant, and whenever one checks the clock, one is surprised at how little time has actually elapsed."

Evidence also suggests that suicidal individuals have a difficult time thinking about the future—which for those who'd use the threat of hell as a deterrent shows just why this strategy isn't likely to be very effective. This temporal narrowing, Baumeister believes, is actually a defensive mechanism that helps the person to withdraw cognitively from thinking about past failures and the anxiety of an intolerable, hopeless future.

Another central aspect of the suicidal person's cognitive deconstruction, says Baumeister, is a dramatic increase in concrete thought. Like the intrusively high self-awareness discussed earlier, this concreteness is often conveyed in suicide notes. Several review articles have noted the relative paucity of "thinking words" in suicide notes, which are abstract, meaningful, high-level terms. Instead, they more often include banal and specific instructions, such as "Don't forget to feed the cat" or "Remember to take care of the electric bill." Real suicide notes are usually suspiciously void of contemplative or metaphysical thoughts, whereas fake suicide notes, written by study participants, tend to include more abstract or high-level terms ("Someday you'll understand how much I loved you" or "Always be happy"). One old study even found that genuine suicide notes contained more references to concrete objects in the environment—physical things— than did "fake" (simulated) suicide notes.

What this cognitive shift to concrete thinking reflects, suggests Baumeister, is the brain's attempt to slip into idle mental labor, thereby avoiding the suffocating feelings that we've been describing. Many suicidal college students, for example, exhibit a behavioral pattern of burying them- selves in dull, routine academic busywork in the weeks beforehand, presumably to enter a sort of "emotional dead- ness" that is "an end in itself." When I was a suicidal adolescent, I remember reading voraciously; it didn't matter what it was that I read—mostly junk novels, in fact—since it was only to replace my own thoughts with those of the writer's. For the suicidal, other people's words can be pulled over one's exhausting ruminations like a seamless glove being stretched over a distractingly scarred hand.

Even the grim, tedious details of organizing one's own suicide can offer a welcome reprieve: "When preparing for suicide, one can finally cease to worry about the future, for one has effectively decided that there will be no future. The past, too, has ceased to matter, for it is nearly ended and will no longer cause grief, worry, or anxiety. And the imminence of death may help focus the mind on the immediate present."

Step 6: Disinhibition. We've now set the mental stage, but it is of course the final act that separates suicidal ideation from an actual suicide. Baumeister speculates that behavioral disinhibition, which is required to overcome the intrinsic fear of causing oneself pain through death, not to mention the anticipated suffering of loved ones left behind to grieve, is another consequence of cognitive deconstruction. This is because it disallows the high-level abstractions (reflecting on the inherent "wrongness" of suicide, how others will feel, even concerns about self-preservation) that, under normal conditions, keep us alive.

A theoretical analysis by the psychiatrist Kimberly Van Orden and her colleagues sheds some additional light on this component of behavioral disinhibition. These authors point out that while there is a considerable number of people who want to kill themselves, suicide itself remains relatively rare. This is largely because, in addition to suicidal desire, the individual needs the "acquired capability for suicide," which involves both a lowered fear of death and an increased physical pain tolerance. Suicide hurts, literally. One acquires this capability, according to these authors' model, by being exposed to related conditions that systematically habituate the individual to physical pain. For example, one of the best predictors of suicide is a nonlethal prior suicide attempt.

But a history of other fear-inducing, physically painful experiences also places one at risk. Physical or sexual abuse as a child, combat exposure, and domestic abuse can also "prep" the individual for the physical pain associated with suicidal behavior. In addition, heritable variants of impulsivity, fearlessness, and greater physical pain tolerance may help to explain why being suicidal often runs in families. Van Orden and her coauthors also cite some intriguing evidence that habituation to pain is not so much generalized to just any old suicide method as often specific to the particular method used to end one's own life. For example, a study on suicides in the U.S. military branches found that guns were most frequently associated with army personnel suicides, hanging and knots for those in the navy, and falling and heights for those in the air force.

So there you have it. It's really not a pretty picture. But again, I do hope that if you ever are unfortunate enough to experience these cognitive dynamics in your own mind—and I, for one, very much have—or if you suspect you're seeing behaviors in others that indicate these thought patterns may be occurring, this information helps you to meta-cognitively puncture suicidal ideation. If there is one thing I've learned since those very dark days of my suicidal years, it's that scientific knowledge changes perspective. And perspective changes everything. *Everything*.

And, as I alluded to at the start, always remember: you're going to die soon enough anyway; even if it's a hundred years from now, that's still the blink of a cosmic eye. In the meantime, live like a scientist—even a controversial one with only a colleague or two in all the world—and treat life as a grand experiment, blood, sweat, tears, and all. Bear in mind that there's no such thing as a failed experiment—only data.

"Scientists Say Free Will Probably Doesn't Exist, Urge 'Don't Stop Believing!' "

Suspend disbelief for a moment and imagine that you have agreed, as a secret agent in some confidential military operation, to travel back in time to the year 1894. To your astonishment, it's a success! And now—after wiping away the magical time-traveling dust from your eyes—you find yourself on the fringes of some Bavarian village, hidden in a camouflaging thicket of wilderness against the edge of town, the distant, disembodied voices of nineteenth-century Germans mingling atmospherically with the unmistakable sounds of church bells.

Quickly you survey your surroundings: you seem to be directly behind a set of old row houses; white linens have been hung out to dry; a little stream tinkles behind you; windows have been opened to let in the warm springtime air. How quaint. No one else appears to be about, although occasionally you glimpse a pedestrian passing between the narrow gaps separating the houses. And then you notice him. There's a quiet, solemn-looking little boy nearby, playing quietly with some toys in the dirt. He looks to be about five years old—a mere kindergartner, in the modern era. It's then that you're reminded of your mission: This is the town of Passau in southern Germany. And that's no ordinary little boy. It's none other than young Adolf Hitler. *What would you do next?*

This scenario is, rather unfortunately for us, in the realm of science fiction. But how you answer the hypothetical question—and others like it—is a matter for psychological scientists, because among other things it betrays your underlying assumptions about whether Hitler, as well as the decisions he made later in his life, was simply the product of his environment acting on his genes or whether he could have acted differently by exerting his "free will." Most scientists in this area aren't terribly concerned over whether free will does or does not exist; instead, they focus on how people's everyday reasoning about free will, particularly in the moral domain, influences their social behaviors and attitudes.

We've already met one of the leading investigators in this area, Roy Baumeister, who so effectively disarmed the psychology underlying suicidal thoughts. Here's Baumeister's take on the psychology of free will:

> At the core of the question of free will is a debate about the psychological causes of action. That is, is the person an autonomous entity who genuinely chooses how to act from among multiple possible options? Or is the person essentially just one link in a causal chain, so that the person's actions are merely the inevitable product of lawful causes stemming from prior events, and no one ever could have acted differently than he or she actually did? . . .
>
> To discuss free will in terms of scientific psychology is therefore to invoke notions of self-regulation, controlled processes, behavioral plasticity, and conscious decision-making.

With this understanding of what psychologists study when they turn their attention to people's beliefs in free will, let's return to the Hitler example above. In your role of this time-traveling secret agent from the twenty-first century, you've been equipped with the following pieces of information. First, the time-traveling technology is still in its infancy, and researchers are doubtful that it will ever succeed again. Second, you have only ten minutes before being zapped back into the present (and two of those minutes have already elapsed since you arrived). Third, you've been informed that seven minutes is just enough time to throttle a five-year-old with your bare hands and to confirm, without a doubt, that the child is dead. This means that you have only one minute left to decide whether or not to kill the little boy.

But you have other options. Seven minutes is also enough time, you've been told by your advisers, to walk into the Hitler residence and hand-deliver to Alois and Klara, Adolf's humorless father and kindly, retiring mother, a specially prepared package of historical documents related to the Holocaust, including clear photographs of their son as a mustachioed führer and a detailed look at the Third Reich four decades later. Nobody knows precisely what effect this would have, but most modern scholars believe that this horrifying preview of World War II would meaningfully alter Adolf's childhood. Perhaps Klara would finally leave her domineering, abusive husband; Alois, unhappy with the idea of his surname becoming synonymous with all that is evil, might change his ways and become a kinder parent; or they might both sit down together with the young Adolf and share with him disturbing death-camp images and testimonies from Holocaust survivors that are so shocking and terrifying that even Adolf himself would come to disdain his much-hated adult persona. But can Adolf really change the course of his life? Does he have free will? Do any of us?

One of the most striking findings to emerge recently in the science of free will is that when people believe—or are led to believe—that free will is just an illusion, they tend to become more antisocial. We'll get back to little Adolf shortly. (Which do you think is the antisocial decision here, to kill or not to kill the Hitler boy?) But before making your decision, have a look at what the science says. The first study to directly demonstrate the antisocial consequences of deterministic beliefs was done by the psychologists Kathleen Vohs and Jonathan Schooler. In their report in *Psychological Science*, Vohs and Schooler invited thirty undergraduates into their lab to participate in what was ostensibly a study about mental arithmetic in which they were asked to calculate the answers to twenty math problems (for example, $1 + 8 + 18 - 12 + 19 - 7 + 17 - 2 + 8 - 4 = ?$) in their heads. But, as social psychology experiments often go, testing something as trivial as the students' math skills was not the real purpose of the study.

Prior to taking the math test, half the group (fifteen participants) were asked to read the following passage from Francis Crick's book *The Astonishing Hypothesis*:

> "You," your joys and your sorrows, your memories and your ambitions, your sense of personal identity and free will, are in fact no more than the behavior of a vast assembly of nerve cells and their associated molecules. Who you are is nothing but a pack of neurons … although we appear to have free will, in fact, our choices have already been predetermined for us and we cannot change that.

In contrast, the other fifteen participants read a different passage from the same book, but one in which Crick makes no mention of free will. And, rather amazingly, when given the opportunity, this second group of people cheated significantly less on the math test than those who read Crick's free-will-as-illusion passage above. (The study was cleverly rigged to measure cheating: participants were led to believe that there was a "glitch" in the computer program, and that if the answer appeared on the screen before they finished the problem, they should hit the space bar and finish the test honestly. The number of space bar clicks throughout the task therefore indicated how honest they were being.) These general effects were replicated in a second experiment using a money allocation task in which participants who were randomly assigned to a *determinism* condition and asked to read statements such as "A belief in free will contradicts the known fact that the universe is governed by lawful principles of science" essentially stole more money than those who'd been randomly assigned to read statements from a *free-will* condition (for example, "Avoiding temptation requires that I exert my free will") or a *neutral* condition with control statements (for example, "Sugarcane and sugar beets are grown in 112 countries").

Vohs and Schooler's findings reveal a rather strange dilemma facing social scientists: If a deterministic understanding of human behavior encourages antisocial behavior, how can we scientists justify communicating our deterministic research findings? In fact, there's a rather shocking line in this *Psychological Science* article, one that I nearly overlooked on my first pass. Vohs and Schooler write: "If exposure to deterministic messages increases the likelihood of unethical actions, then identifying approaches for insulating the public against this danger becomes imperative."

Perhaps you missed it on your first reading too, but the authors are making an extraordinary suggestion. They seem to be claiming that the public "can't handle the truth" and that we should somehow be protecting them (lying to them?) about the true causes of human social behaviors. Perhaps they're right. Consider the following example.

A middle-aged man hires a prostitute, knowingly exposing his wife to a sexually transmitted infection and exploiting a young drug addict for his own pleasure. Should the man be punished somehow for his transgression? Should we hold him accountable? Most people, I'd wager, wouldn't hesitate to say yes to both questions.

But what if you thought about it in the following slightly different, scientific terms? The man's decision to have sex with this woman was in accordance with his physiology at that time, which had arisen as a consequence of his unique developmental experiences, which occurred within a particular cultural environment in interaction with a particular genotype, which he inherited from his particular parents, who inherited genetic variants of similar traits from their own particular parents, ad infinitum. Even his ability to inhibit or "override" these forces, or to understand his own behavior, is the product itself of these forces! What's more, this man's brain acted without first consulting his self-consciousness; rather, his neurocognitive system enacted evolved behavioral algorithms that responded, either normally or in error, in ways that had favored genetic success in the ancestral past.

Given the combination of these deterministic factors, could the man have responded any other way to the stimuli he was confronted with? Attributing personal responsibility to this sap becomes merely a social convention that reflects only a naive understanding of the causes of his behaviors. Like us judging him, this man's self merely plays the role of spectator in his body's sexual affairs. There is only the embodiment of a man who is helpless to act in any way that is contrary to his particular nature, which is a derivative of a more general nature. The self is only a deluded creature that thinks it is participating in a moral game when in fact it is just an emotionally invested audience member.

If this deterministic understanding of the man's behaviors leads you to feel even a smidgen more sympathy for him than you otherwise might have, that reaction is precisely what Vohs and Schooler are warning us about. How can we fault this "pack of neurons"—let alone punish him—for acting as his nature dictates, even if our own nature would have steered us otherwise? What's more, shouldn't we be more sympathetic toward our own moral shortcomings? After all, we can't help who we are either. Right?

In fact, a study in *Personality and Social Psychology Bulletin* by Baumeister and his colleagues found that simply exposing people to deterministic statements, such as "Like everything else in the universe, all human actions follow from prior events and ultimately can be understood in terms of the movement of molecules," made them act more aggressively and selfishly compared with those who read statements endorsing the idea of free will, such as "I demonstrate my free will every day when I make decisions," or those who simply read neutral statements, such as "Oceans cover 71 percent of the earth's surface." Participants who'd been randomly assigned to the deterministic condition, for example, were less likely than those from the other two groups to give money to a homeless person or to allow a classmate to use their cellular phone. In discussing the societal implications of these results,

Baumeister and his coauthors echo Vohs and Schooler's concerns about "insulating the public" from a detailed understanding of the causes of human social behaviors: "Some philosophical analyses may conclude that a fatalistic determinism is compatible with highly ethical behavior, but the present results suggest that many lay-persons do not yet appreciate that possibility."

These laboratory findings that demonstrate the anti-social consequences of viewing individual human beings as hapless pinballs trapped in a mechanical system—even when, in point of fact, that's pretty much what we are—are enough to give me pause in my scientific proselytizing. Returning to innocent little Adolf, we could, of course, play with this particular example forever. It's an unpalatable thought, but what if one of the children slaughtered at Auschwitz would have grown up to be even more despised than Hitler, as an adult ordering the deaths of ten million? Isn't your ability to make a decision a question fundamentally about your own free will? And so on. But the point is not to play the what-if Hitler game in some infinite regress but rather to provoke your intuitions about free will without asking you directly whether you believe in it or not. As any good scientist knows, what people say they believe doesn't always capture their private psychology.

In this case, it's not so much your decision to kill the child or to deliver the package to his parents that research psychologists would be interested in. Rather, it's how you would *justify* your decision (for example, "I'd kill him because [fill in the blank here]" or "I'd deliver the package because [fill in the blank]") that would illuminate your thinking about Hitler's free will. On the face of it, strangling an innocent five-year-old seems rather anti-social, and so perhaps hearing a deterministic message before answering this question would lead you to kill him (for instance, "Hitler is evil, he will grow up to murder people no matter what—he has no free will to do otherwise"). For some people, however, the decision not to kill the innocent boy is the antisocial one, because it may well mean the unthinkable for more than six million fellow human beings.

I, for one, wouldn't hesitate to gleefully strangle that little prick in 1894 Passau. (The fact that I recently visited Auschwitz may have something to do with that.) I can't help but *feel* that Hitler could have raised his hand at any time and quashed the so-called Final Solution of the Jewish people before it ever began. This justification seems to reveal my hidden belief in free will: Adolf could have acted differently but chose not to. That is to say, the chain of causal events preceding Hitler's rise to power seems largely irrelevant to me, or at least inconsequential. His bad deeds would have occurred irrespective of the vicissitudes of his personal past. There is something essentially evil about this individual. And so I decide to kill the child: it's probably best in this instance, I seem to be saying, to slay the beast while it's still lying dormant in a little boy playing with toy soldiers.

But you might opt for a less homicidal way to spend your time with Adolf. For example, if you spare the life of this pasty, forlorn kid and decide to deliver the package to his parents because, you say, had the Hitlers known what was to become of their troubled son, they would have raised him otherwise, and this change in his early environment would almost certainly have prevented genocide, this entails that you subscribe more to the principle of causal determinism.

In any event, your minute is up! So what's it going to be—and *why*? With millions of future lives at stake, do you murder the innocent five-year-old boy as a preemptive homicide? Do you deliver the package to his parents, in the hopes that the shocking vision of the Holocaust will lead Adolf—one way or another—to choose a different career path, or even to flub his own rise to fame from all the pressure? Or, like those who lived in Nazi Germany and who were bombarded with (false) deterministic messages about the Jews, do you simply not intervene at all?

The Rat That Wouldn't Stop Laughing: Joy and Mirth in the Animal Kingdom

Once, while in a drowsy, altitude-induced delirium thirty-five thousand feet somewhere over Iceland, I groped mindlessly for the cozy blue blanket poking out beneath my seat, only to realize—to my unutterable horror—that I was in fact tugging soundly on a wriggling, sock-covered big toe. Now, with a temperament such as mine, life tends to be one awkward conversation after the next, so when I turned around, smiling, to apologize to the owner of this toe, my gaze was met by a very large man whose grunt suggested that he was having some difficulty in finding the humor in this incident.

Unpleasant, sure, but I now call this event serendipitous. As I rested my head back against that sanitary-paper-covered airline pillow, my midflight mind lit away to a much happier memory, one involving another big toe, yet this one belonging to a noticeably more good-humored animal than the one sitting behind me. This other toe—which felt every bit as much as its overstuffed human equivalent did, I should add—was attached to a 450-pound western lowland gorilla, with calcified gums, named King. When I was twenty, and he was twenty-seven, I spent much of the summer of 1996 with my toothless friend King, listening to Frank Sinatra and the Three Tenors, playing chase from one side of his exhibit to the other, and tickling his toes. He'd lean back in his night house, stick out one huge ashen-gray foot through the bars of his cage, and leave it dangling there in anticipation, erupting in shoulder-heaving guttural laughter as I'd grab hold of one of his toes and gently give it a palpable squeeze. He almost couldn't control himself when, one day, I leaned down to act as though I were going to bite on that plump digit. If you've never seen a gorilla in a fit of laughter, I'd recommend searching out such a sight before you pass from this world. It's something that would stir up cognitive dissonance in even the heartiest of creationists.

Do animals other than humans have a sense of humor? Perhaps in some ways, yes. But in other ways there are likely uniquely human properties to such emotions. Aside from anecdotes, we know very little about nonhuman primate laughter and humor, but some of the most significant findings to emerge in comparative science over the past decade have involved the unexpected discovery that rats—particularly juvenile rats—laugh. That's right: rats laugh. At least, that's the unflinching argument being made by researcher Jaak Panksepp, who published a remarkable, and rather heated, position paper on the subject in *Behavioural Brain Research*.

In particular, Panksepp's work has focused on "the possibility that our most commonly used animal subjects, laboratory rodents, may have social-joy type experiences during their playful activities and that an important communicative-affective component of that process, which invigorates social engagement, is a primordial form of laughter." Now, before you go imagining some chortling along the lines of one rakish Stuart Little (or was he a mouse?), real rat laughter doesn't tend to sound very much like the human variety, which normally involves pulsating sound bursts starting with a vocalized inhalation and consisting of a series of short distinct trills with almost isochronous time intervals. The stereotypical sound of human laughter is an aspirated *h* followed by a vowel, usually *a*, and owing largely to our larynx, is rich in harmonics. By contrast, rat laughter comes in the form of high-frequency 50 kHz ultrasonic calls, or "chirps," that are distinct from other vocal emissions in rats. Here's how Panksepp describes his discovery of the phenomenon:

> Having just concluded perhaps the first formal (i.e., well-controlled) ethological analysis of rough-and-tumble play in the human species in the late 1990s, where laughter was an abundant response, I had the "insight" (perhaps delusion) that our 50 kHz chirping response in playing rats might have some ancestral relationship to human laughter. The morning after, I came to the lab and asked my undergraduate assistant at the time to "come tickle some rats with me."

Over the ensuing years, Panksepp and his research assistants systematically conducted study after study on rat laughter, revealing a striking overlap between the functional and expressive characteristics of this chirping response in young rodents and laughter in young human children. To elicit laughter in his rat pups, Panksepp used a technique that he called "heterospecific hand play," which is essentially just jargon for tickling. "For this maneuver to work well," he writes,

> one must be adept at performing dynamic forms of interspecies interactions. With some modest training, most investigators can readily acquire the skill—it is

rather similar to the dynamic hand and finger movements that one might use in tickling young human children, who can be provoked into flurries of playfulness and peals of laughter by this simple maneuver.

Rats are particularly ticklish, it seems, in their nape area, which is also where juveniles target their own play activities such as pinning behavior. Panksepp soon found that the most ticklish rats—which, empirically, means simply those rats that emitted the most frequent, robust, and reliable 50 kHz chirps in human hands—were also the most naturally playful individuals among the rat subjects. And he discovered that inducing laughter in young rats promoted bonding: tickled rats would actively seek out specific human hands that had made them laugh previously. In addition, and as would be expected in humans, certain aversive environmental stimuli dramatically reduced the occurrence of laughter among rodent subjects. For example, even when tickling stimulation was kept constant, chirping diminished significantly when the rat pups got a whiff of cat odor, when they were very hungry, or when they were exposed to unpleasant bright lights during tickling. Panksepp also discovered that adult females were more receptive to tickling than males, but in general it was difficult to induce tickling in adult animals "unless they have been tickled abundantly when young." Finally, when rat pups were given the choice between two different adults—one that still spontaneously chirped a lot and one that didn't—they spent substantially more time with the apparently happier grown-up rat.

Perhaps not surprisingly, Panksepp has encountered an unfortunate resistance to his interpretation of this body of findings, especially among his scientific colleagues. Yet he protests:

We have tried to negate our view over and over, and have failed to do so. Accordingly, we feel justified in cautiously advancing and empirically cultivating the theoretical possibility that there is some kind of an ancestral relationship between the playful chirps of juvenile rats and infantile human laughter. This hypothesis has caused great consternation for many colleagues in the behavioral neuroscience community; they see no reason for anyone to go so far out on the ontological limb. Several colleagues have discouraged this kind of theorizing, suggesting that this is fundamentally inappropriate, even embarrassing, for members of our discipline to speak about animal brain functions in such blatantly anthropomorphic ways.

Now, Panksepp would be the first to acknowledge that his findings do not imply that rats have a "sense of humor," only that there appear to be evolutionary contiguities between laughter in human children during rough-and-tumble play and the expression of similar vocalizations in young rats. A sense of humor—especially adult humor—requires cognitive mechanisms that may or may not be present in other species. He does suggest, however, that this may be an empirically falsifiable hypothesis: "If a cat . . . had been a persistently troublesome feature of a rat's life, might that rat show a few happy chirps if something bad happened to its nemesis? Would a rat chirp if the cat fell into a trap, or was whisked up into the air by its tail? We would not recommend such mean-spirited experiments to be conducted, but would encourage anyone who wishes to go in that direction to find more benign ways to evaluate those issues."

Differences between laughing "systems" among mammals are reflected by cross-species structural differences in brain regions as well as in vocal architecture. In the same issue of *Behavioural Brain Research*, the neuropsychologist Martin Meyer and his colleagues describe these differences in rich detail. For example, although brain-imaging studies of human participants watching funny cartoons or listening to jokes reveal the activation of evolutionarily ancient structures such as the amygdala and nucleus accumbens, more recently evolved, "higher-order" structures are also activated, including distributed regions of the frontal cortex. So although nonhuman primates laugh—in fact, the authors describe how in 1943 a team of vivisectionists discovered that when they stimulated the diencephalon, midbrain pons, and medulla of macaque monkeys, the animals started laughing uncontrollably and displaying play faces—human humor seems also to involve more specialized cognitive networks that are unshared by other species.

Laughter in our own species, of course, is triggered by a range of social stimuli and occurs under a wide umbrella of emotions, not always positive. To name just a few typical emotional contexts for laughter, it can accompany joy, affection, amusement, cheerfulness, surprise, nervousness, sadness, fear, shame, aggression, triumph, taunt, and schadenfreude (pleasure in another's misfortune). In fact, like masturbatory fantasies, laughter can occur even in the physical absence of any social stimuli. If you've ever noticed someone walking down the sidewalk, head down, smiling, suppressing an embarrassing chuckle for fear that it might falsely signal schizophrenia to a naive onlooker, this person is actually engaged in a fairly sophisticated cognitive activity, one where she's "re-presenting" a real or imagined humorous scene in her mind's eye.

But typically, laughter serves as a rich social signal and occurs in the presence of others. This phenomenon led the psychologist Diana Szameitat and her team to explore the possible adaptive function of human laughter. Their study, published in *Emotion*, provides the first experimental evidence demonstrating that human beings possess an uncanny ability to detect a laugher's psychological intent by the phonetic qualities of laugh sounds alone. And sometimes, the authors point out, laughter signals some very aggressive intentions, a fact that should—from an evolutionary perspective—motivate appropriate, or biologically adaptive, behavioral responses on the part of the listener.

Now, it's difficult, if not impossible, to induce genuine, discrete emotions under controlled laboratory conditions, so for their first study Szameitat and her colleagues did the next best thing: they hired eight professional actors (three men and five women) and recorded them laughing. This isn't ideal, obviously, and the researchers acknowledge the limited applicability of using "emotional portrayals" rather than genuine emotions. But they did use auto-induction techniques, instructing the actors to get into full character by using their imaginations, bodily movements, and emotional recall. In other words, "the actors were instructed to focus exclusively on the experience of the emotional state but not at all on the outward expression of the laughter." Here are the four basic laughing types that the actors were asked to perform, along with the sample descriptions and scenarios used to facilitate the actors' getting into character for their roles:

1. *Joyful laughter*: Meeting a good friend after not having seen him for a very long time.
2. *Taunting laughter*: Laughing at an opponent after having defeated him. It reflects the emotion of sneering contempt and serves to humiliate the listener.

3. *Schadenfreude laughter*: Laughing at another person to whom a misfortune has happened, such as slipping in dog dirt. As opposed to taunting laughter, however, the laugher doesn't want to seriously harm the other person.

4. *Tickling laughter*: Laughing when being physically, literally, tickled.

Once these recordings were collected, seventy-two English-speaking participants were invited to the laboratory, given a set of headphones, and instructed to identify the emotions behind these laughter sequences. These people listened to a lot of laugh sequences—429 laugh tracks total, each representing a randomly interspersed laugh pulse ranging from three to nine seconds in length, so that there were 102–111 laughs per emotion. (This took them about an hour—a nightmarish scenario reminding me of 1980s television sitcoms and focusing my attention on the peculiar laugh tracks in the background.) But the findings were impressive; the participants were able to correctly classify these laugh tracks by their often subtly expressed emotions with a success rate significantly above chance.

In a second study, the procedure was nearly identical, but participants were asked a different set of questions concerning the social dynamics. Specifically, for each laugh track, they were asked if the "sender" (that is, the laugher) was in a physically excited or a calm state, whether he or she was dominant or submissive relative to the "receiver" (that is, the subject of the laugh), in a pleasant or unpleasant state, and whether he or she was being friendly or aggressive toward the receiver. For this second study, there was no "correct" or "incorrect" response, since perceiving these characteristics in the laugh tracks involved subjective attributions. Yet, as predicted, each category of laughter (joy, taunt, schadenfreude, tickling) had a unique profile on these four social dimensions. That is to say, the participants used these sounds to reliably infer specific social information regarding the unseen situation. Joy, for example, invoked judgments of low arousal, submissiveness, and positive valence on both sides. Taunting laughter clearly stood out: it was very dominant and was the only sound that was perceived by the participants as having a negative valence directed at the receiver.

The participants' perception of schadenfreude laughter was especially interesting. It was heard as being dominant, but not quite so dominant as taunt; senders who engaged in such laughter were judged as being in a positive state, more so than taunt, but less than tickle; and schadenfreude laughter was heard as being neither aggressive nor friendly toward the receiver, but neutral. According to the authors, whose interpretations of these data again were inspired by evolutionary logic: "Schadenfreude laughter might therefore represent a precise (and socially tolerated) tool to dominate the listener without concurrently segregating him from group context."

In any event, I'd like to think that I was witnessing pure, unadulterated joy in King those many years ago, but of course my brain isn't made to decipher distinct emotive states in gorillas. He's since been laughing, apparently, at Ellen DeGeneres while watching her on television in his cage; two is a small sample size, I realize, but perhaps he finds homosexual human beings particularly comical. Yet it brings me joy to think of the evolution of joy. And I've got to say, those rat data have me seriously considering a return to my old vegetarian days—not that I dine on rats, of course, but laughing animals do make the prospect of animal suffering unusually salient and uncomfortable in my mind.

If only dead pigs weren't so spectacularly delicious.

Notes

Acknowledgments

Index

Notes

How Are They Hanging? This Is Why They Are

21   *"With the possible exception"*: Gordon G. Gallup Jr., Mary M. Finn, and Becky Sammis, "On the Origin of Descended Scrotal Testicles: The Activation Hypothesis," *Evolutionary Psychology* 7, no. 4 (2009): 519.

21   *"Not only is the skin"*: Ibid., 519.

23   *According to a 2009 report*: Stany W. Lobo et al., "Asymmetric Testicular Levels in the Crotch: A Thermodynamic Perspective," *Medical Hypotheses* 72, no. 6 (2009): 759–60.

24   *"In our view"*: Gallup, Finn, and Sammis, "On the Origin of Descended Scrotal Testicles": 521.

25   *"Any account of descended"*: Ibid., 523.

26   *Or to think about it another way*: That's not to say that such individuals don't exist. Cases of algolagnia (from the Greek *algos* [pain] and *lagneia* [lust]) do exist, and some of these people derive their primary sexual satisfaction from insults to their erogenous zones. But this is so bizarre that many contemporary researchers believe that algolagnia—*especially* when one can only get

aroused by testicular pain or vaginal tearing—can only be understood as signaling a hazardous neurological disorder involving miscoding noxious stimuli.

So Close, and Yet So Far Away: The Contorted History of Autofellatio

29   "*a considerable portion*": Alfred C. Kinsey, Wardell B. Pomeroy, and Clyde E. Martin, *Sexual Behavior in the Human Male* (Philadelphia: W. B. Saunders, 1948), 510.

29   *had a bone removed*: Grazia D'Annunzio, "The Randy Dandy," *New York Times*, www.nytimes.com/2009/09/13/style/tmagazine/13slijperw.html.

29   a "*very disturbed*" *patient*: Frances Millican et al., "Oral Autoaggressive Behavior and Oral Fixation," in *Masturbation: From Infancy to Senescence*, ed. Irwin M. Marcus and John J. Francis (Madison, Conn.: International Universities Press, 1975), 150.

29-30 *lonely twenty-two-year-old serviceman*: Jesse O. Cavenar, Jean G. Spaulding, and Nancy T. Butts, "Autofellatio: A Power and Dependency Conflict," *Journal of Nervous and Mental Disease* 165, no. 5 (1977): 356–60.

30   *typical jargon-filled language*: Frank Orland, "Factors in Autofellatio Formation," *International Journal of Psychoanalysis* 52, no. 3 (1971): 289–96.

30   *The very first published psychiatric case*: Eugen Kahn and Ernest G. Lion, "A Clinical Note on a Self-Fellator," *American Journal of Psychiatry* 95, no. 1 (1938): 131–33.

32   *a theme beginning to emerge*: William Guy and Michael H. Finn, "A Review of Auto-Fellatio: A Psychological Study of Two New Cases," *Psychoanalytic Review* 41, no. 4 (1954): 354–58.

33   *virginal staff sergeant*: Morris M. Kessler and George E. Poucher, "AutoFellatio: Report of a Case," *American*

Journal of Psychiatry 103, no. 1 (1946): 94–96.

34 *especially self-sufficient female patient*: Orland, "Factors in Autofellatio Formation."

Why Is the Penis Shaped Like That? The Extended Cut

38 *"a longer penis would"*: Gordon G. Gallup Jr. and Rebecca L. Burch, "Semen Displacement as a Sperm Competition Strategy in Humans," *Evolutionary Psychology* 2, no. 1 (2004): 14.

39 *"Examples include group sex"*: Ibid., 15.

39 *In a series of studies*: Gordon G. Gallup Jr. et al., "The Human Penis as a Semen Displacement Device," Evolution and Human Behavior 24, no. 4 (2003): 277–89.

43 *"Is it possible"*: Gallup and Burch, "Semen Displacement as a Sperm Competition Strategy in Humans": 16.

Not So Fast . . . What's So "Premature" About Premature Ejaculation?

48 *"an expeditious partner who"*: Lawrence K. Hong, "Survival of the Fastest: On the Origin of Premature Ejaculation," *Journal of Sex Research* 20, no. 2 (1984): 113.

49 *"the ancestry of Homo sapiens"*: Ibid., 117.

49 *a 2009 article*: Patrick Jern et al., "Evidence for a Genetic Etiology to Ejaculatory Dysfunction," *International Journal of Impotence Research* 21, no. 1 (2009): 62–67.

50 *Adding further credence*: Patrick Jern et al., "Subjectively Measured Ejaculation Latency Time and Its Association with Different Sexual Activities While Controlling for Age and Relationship Length," *Journal of Sexual Medicine* 6, no. 9 (2009): 2568–78.

51 *"there would be little"*: Ray Bixler, "Of Apes and Men

(Including Females)," *Journal of Sex Research* 22, no. 2 (1986): 265.

An Ode to the Many Evolved Virtues of Human Semen

53   "*Our interest in the*": Rebecca L. Burch and Gordon G. Gallup Jr., "The Psychobiology of Human Semen," in *Female Infidelity and Paternal Uncertainty: Evolutionary Perspectives on Male Anti-cuckoldry Tactics*, ed. Steven M. Platek and Todd K. Shackelford (Cambridge, Mass.: Cambridge University Press, 2006), 141.

54   "*this struck us as peculiar*": Ibid., 141.

56   *The most significant findings*: Gordon G. Gallup Jr., Rebecca L. Burch, and Steven M. Platek, "Does Semen Have Antidepressant Properties?," *Archives of Sexual Behavior* 31, no. 3 (2002): 289–93.

57   *pulsing through one's veins*: And it gets better. A smaller percentage (4.5 percent) of the sexually active women who "never" used condoms were less likely to have attempted suicide than were those who "sometimes" (7.4 percent) and "usually" (28.9 percent) and "always" (13.2 percent) used condoms.

57   "*It is important to*": Ibid: 291.

58   "*The body becomes the*": Dave Holmes and Dan Warner, "The Anatomy of Forbidden Desire: Men, Penetration, and Semen Exchange," *Nursing Inquiry* 12, no. 1 (2005): 18.

59   *make HIV up to*: Jan Münch et al., "Semen-Derived Amyloid Fibrils Drastically Enhance HIV Infection," *Cell* 131, no. 6 (2007): 1059–71.

60   "*by the age of 11–12*": Gilbert Herdt and Martha McClintock, "The Magical Age of 10," *Archives of Sexual Behavior* 29, no. 6 (2000): 596.

61   *What are female hormones doing*: Burch and Gallup Jr., "Psychobiology of Human Semen": 159.

61   *"Thus it would appear"*: Ibid., 160.

The Hair Down There: What Human Pubic Hair Has in Common with Gorilla Fur

66   *"the pubic hair was"*: Samar K. Bhowmick, Tracy Ricke, and Kenneth R. Retig, "Sexual Precocity in a 16-Month-Old Boy Induced by Indirect Topical Exposure to Testosterone," *Clinical Pediatrics* 46, no. 6 (2007): 540–41.

67   *"Although naked apes [humans]"*: Robin A. Weiss, "Apes, Lice, and Prehistory," *Journal of Biology* 8, no. 2 (2009): 20.

69   *"On the basis of "*: Ibid.

70   *Flinders University psychologists*: Marika Tiggemann and Suzanna Hodgson, "The Hairlessness Norm Extended: Reasons for and Predictors of Women's Body Hair Removal at Different Body Sites," *Sex Roles* 59, no. 11–12 (2008): 889–97.

70   *In a separate study*: Marika Tiggemann, Yolanda Martins, and Libby Churchett, "Hair Today, Gone Tomorrow: A Comparison of Body Hair Removal Practices in Gay and Heterosexual Men," *Body Image* 5, no. 3 (2008): 312–16.

Bite Me: The Natural History of Cannibalism

74   *"The point is that"*: Lewis Petrinovich, *The Cannibal Within* (Piscataway, N.J.: Aldine Transaction, 2000), 107.

75   *"After he cut the first toe"*: Gregory M. De Moore and Marcus Clement, "Self-Cannibalism: An Unusual Case of Self-Mutilation," *Australian and New Zealand Journal of Psychiatry* 40, no. 10 (2006): 937.

76   *Osteoarchaeological research at*: Alban Defleur et al., "Neanderthal Cannibalism at Moula-Guercy, Ardèche, France," *Science* 286, no. 5437 (1999): 128–31.

78   *"This sustained heterozygote advantage"*: John

Brookfield, "Human Evolution: A Legacy of Cannibalism in Our Genes?," *Current Biology* 13, no. 15 (2003): 592.

78 *such cases reflect essentialist beliefs*: Bruce Hood, *SuperSense: Why We Believe in the Unbelievable* (New York: HarperOne, 2009).

79 *"There is no form"*: Margaret St. Clair, foreword to *To Serve Man: A Cookbook for People*, by Karl Würf (Philadelphia: Owlswick Press, 1976), 1.

The Human Skin Condition: Acne and the Hairless Ape

80 *dealing with hair-covered flesh*: Stephen Kellett and Paul Gilbert, "Acne: A Biopsychosocial and Evolutionary Perspective with a Focus on Shame," *British Journal of Health Psychology* 6, no. 1 (2001): 1–24.

81 *Consider a scene*: Jean-Paul Sartre, *No Exit: And Three Other Plays* (1946; New York: Vintage, 1989), 21.

82 *"I can feel the"*: Craig Murray and Katherine Rhodes, "The Experience and Meaning of Adult Acne," *British Journal of Health Psychology* 10, no. 2 (2005): 193.

82 *"When I'm talking to"*: Ibid., 192.

83 *"Society doesn't allow"*: Ibid., 196.

83 *these were the results reported*: Tracey A. Grandfield, Andrew R. Thompson, and Graham Turpin, "An Attitudinal Study of Responses to a Range of Dermatological Conditions Using the Implicit Association Test," *Journal of Health Psychology* 10, no. 6 (2005): 821–29.

84 *One-third of New Zealand*: Diana Purvis et al., "Acne, Anxiety, Depression, and Suicide in Teenagers: A Cross-Sectional Survey of New Zealand Secondary School Students," *Journal of Paediatrics and Science Health* 42, no. 12 (2006): 793–96.

84 *"it is our considered"*: Marion Sulzberger and Sadie Zaidens, "Psychogenic Factors in Dermatologic Disorders," *Medical Clinics of North America* 32 (1948): 684.

85 *certain human populations*: Loren Cordain et al., "Acne Vulgaris: A Disease of Western Civilization," *Archives of Dermatology* 138, no. 12 (2002): 1584–90.

Naughty by Nature: When Brain Damage Makes People Very, Very Randy

89 *"The brain is the physical manifestation"*: Shelley Batts, "Brain Lesions and Their Implications in Criminal Responsibility," *Behavioral Sciences and the Law* 27, no. 2 (2009): 267.

91 *"all seven children"*: Sunil Pradhan, Madhurendra N. Singh, and Nirmal Pandey, "Klüver-Bucy Syndrome in Young Children," *Clinical Neurology and Neurosurgery* 100, no. 4 (1998): 256.

92 *"Why don't we"*: Shawn J. Kile et al., "Alzheimer Abnormalities of the Amygdala with Klüver-Bucy Syndrome Symptoms: An Amygdaloid Variant of Alzheimer Disease," *Archives of Neurology* 66, no. 1 (2009): 125.

92 *"was an intelligent"*: D. N. Mendhekar and Harpreet S. Duggal, "Sertraline for Klüver-Bucy Syndrome in an Adolescent," *European Psychiatry* 20, no. 4 (2005): 355.

93 *she began performing fellatio*: John A. Anson and Donald T. Kuhlman, "Post-Ictal Klüver-Bucy Syndrome After Temporal Lobectomy," *Journal of Neurology, Neurosurgery, and Psychiatry* 56, no. 3 (1993): 311–13.

93–4 *"becoming sexually aggressive"*: Vanessa Arnedo, Kimberly Parker-Menzer, and Orrin Devinsky, "Forced Spousal Intercourse After Seizures," *Epilepsy and Behavior* 16, no. 3 (2009): 563.

94 *join him and his wife*: Dietrich Blumer, "Hypersexual

Episodes in Temporal Lobe Epilepsy," *American Journal of Psychiatry* 126, no. 8 (1970): 1099–106.

94   *In 2003, the neurologists*: Jeffrey Burns and Russell Swerdlow, "Right Orbitofrontal Tumor with Pedophilia Symptom and Constructional Apraxia Sign," *Archives of Neurology* 60, no. 3 (2003): 437–40.

94   *In a more recent case*: Julie Devinsky, Oliver Sacks, and Orrin Devinsky, "Klüver-Bucy Syndrome, Hypersexuality, and the Law," *Neurocase: The Neural Basis of Cognition* 16, no. 2 (2009): 140–45.

How the Brain Got Its Buttocks: Medieval Mischief in Neuroanatomy

98   *In their first article*: Régis Olry and Duane Haines, "Fornix and Gyrus Fornicatus: Carnal Sins?," *Journal of the History of the Neurosciences* 6, no. 3 (1997): 338–39.

98   *"The real etymology of "*: Ibid., 338.

99   *In a follow-up article*: Régis Olry and Duane Haines, "The Brain in Its Birthday Suit: No More Reason to Be Ashamed," *Journal of the History of the Neurosciences* 17, no. 4 (2008): 461–64.

Lascivious Zombies: Sex, Sleepwalking, Nocturnal Genitals—and You

102 *And thank goodness*: Carlos H. Schenck, Isabelle Arnulf, and Mark W. Mahowald, "Sleep and Sex: What Can Go Wrong? A Review of the Literature on Sleep Related Disorders and Abnormal Sexual Behaviors and Experiences," *Sleep* 30, no. 6 (2007): 683–702.

103 *Consider the case of*: Peter B. Fenwick, "Sleep and Sexual Offending," *Medicine, Science, and the Law* 36, no. 2 (1996): 122–34.

105 *In a 2007 issue*: Monica L. Andersen et al., "Sexsomnia: Abnormal Sexual Behavior During Sleep," *Brain Research Reviews* 56, no. 2 (2007): 271–82.

105 *In a 1996 issue*: Fenwick, "Sleep and Sexual Offending."

106 *"Some time later"*: Mia Zaharna, Kumar Budur, and Stephen Noffsinger, "Sexual Behavior During Sleep: Convenient Alibi or Parasomnia," *Current Psychiatry* 7, no. 7 (2008): 21.

107 *"An automatism is an"*: Fenwick, "Sleep and Sexual Offending," 131.

107 *the London sleep researcher*: Irshaad Osman Ebrahim, "Somnambulistic Sexual Behavior (Sexsomnia)," *Journal of Clinical Forensic Medicine* 13, no. 4 (2006): 219–24.

108 *After five years of waking up*: Schenck, Arnulf, and Mahowald, "Sleep and Sex."

Humans Are Special and Unique: We Masturbate. A Lot

111 *"on the majority of occasions"*: R. Robin Baker and Mark A. Bellis, "Human Sperm Competition: Ejaculate Adjustment by Males and the Function of Masturbation," *Animal Behavior* 46, no. 5 (1993): 871.

111 *"The advantage to the male"*: Ibid., 863.

112 *"The flowback emerges"*: Ibid., 864.

113 *"scourge of the human race"*: Jeffrey Jensen Arnett, "G. Stanley Hall's Adolescence: Brilliance and Nonsense," *History of Psychology* 9, no. 3 (2006): 192.

114 *In the early 1980s, scientists*: Simon J. Wallis, "Sexual Behavior and Reproduction of Cercocebus albigena johnstonii in Kibale Forest, Western Uganda," *International Journal of Primatology* 4, no. 2 (1983): 153–66.

114 *"During each observation"*: E. D. Starin, "Masturbation Observations in Temminck's Red Colobus," *Folia Primatologica* 75, no. 2 (2004): 115.

115 *"The possibility that the types"*: Gilbert Van Tassel Hamilton, "A Study of Sexual Tendencies in Monkeys

and Baboons," *Journal of Animal Behavior* 4, no. 5 (1914): 296.

115 *"Of all my male monkeys"*: Ibid., 314.

116 *"Jimmy promptly endeavoured"*: Ibid., 315.

117 *"a sort of intoxication"*: Wilhelm Stekel, *Auto-Erotism: A Psychiatric Study of Onanism and Neurosis* (New York: Grove Press, 1961), 139.

118 *"I see in front of me"*: Narcyz Lukianowicz, "Imaginary Sexual Partner: Visual Masturbatory Fantasies," *Archives of General Psychiatry* 3, no. 4 (1960): 438.

118 *"In them he 'saw'"*: Ibid., 441.

120 *In a 1990 study*: Bruce J. Ellis and Donald Symons, "Sex Differences in Sexual Fantasy: An Evolutionary Psychological Approach," *Journal of Sex Research* 27, no. 4 (1990): 527–55.

120 *In their review of research findings*: Harold Leitenberg and Kris Henning, "Sexual Fantasy," *Psychological Bulletin* 117, no. 3 (1995): 469–96.

122 *"Because people who are deprived"*: Ibid., 477.

Pedophiles, Hebephiles, and Ephebophiles, Oh My: Erotic Age Orientation

128 *"Between the age limits"*: Vladimir Nabokov, Lolita (1955; New York: Random House, 1997), 16.

130 *"You are watching"*: Ray Blanchard et al., "Pedophilia, Hebephilia, and the DSM-V," *Archives of Sexual Behavior* 38, no. 3 (2009): 339.

131 *"Imagine how much"*: Thomas K. Zander, "Adult Sexual Attraction to Early-Stage Adolescents: Phallometry Doesn't Equal Pathology," *Archives of Sexual Behavior* 38, no. 3 (2008): 329.

132 *"alliance formation theory"*: Frank Muscarella, "The Evolution of Homoerotic Behavior in Humans," *Journal of Homosexuality* 40, no. 1 (2000): 51–77.

133 *"as there was between"*: Oscar Wilde, "The Love That Dare Not Speak Its Name," www.phrases.org.uk/

meanings/the-love-that-dare-not-speak-its-name.html.

134 *The push to pathologize*: Karen Franklin, "The Public Policy Implications of 'Hebephilia': A Response to Blanchard et al.," *Archives of Sexual Behavior* 38, no. 3 (2008): 319–20.

134 *"a textbook example of"*: Ibid., 319.

135 *"Wilde took a key"*: André Gide, *If It Die: An Autobiography* (New York: Random House, 1935), 288.

136 *"judged the greatest"*: "André Gide Is Dead: Noted Novelist, 81," www.andregide.org/remembrance/nytgide.html.

136 *"fleshy, full-lipped, languorous young boys"*: Posner, Donald. "Caravaggio's Homo-Erotic Early Works," *Art Quarterly* 34 (1971): 301–324.

Animal Lovers: Zoophiles Make Scientists Rethink Human Sexuality

141 *"To a considerable extent"*: Alfred C. Kinsey, Wardell B. Pomeroy, and Clyde E. Martin, *Sexual Behavior in the Human Male* (Philadelphia: W. B. Saunders, 1948), 675–76.

141 *The first case study*: Christopher M. Earls and Martin L. Lalumière, "A Case Study of Preferential Bestiality (Zoophilia)," *Sexual Abuse* 14, no. 1 (2002): 83–88.

142 *"As I grew into adolescence"*: Christopher M. Earls and Martin L. Lalumière, "A Case Study of Preferential Bestiality," *Archives of Sexual Behavior* 38, no. 4 (2009): 606.

144 *"When that black mare"*: Ibid., 606.

144 *Another pioneering researcher*: Hani Miletski, *Understanding Bestiality and Zoophilia* (Bethesda, Md.: self-published, 2002).

145 *"The vehemence with which"*: Peter Singer, "Heavy Petting," *Nerve,*www.utilitarian.net/singer/by/2001.htm.

146 *"Apparently uncertain as to"*: Rebecca Cassidy,

"Zoosex and Other Relationships with Animals," in *Transgressive Sex: Subversion and Control in Erotic Encounters*, ed. Hastings Donnan and Fiona Magowan (New York: Berghahn Press, 2009), p. 95.

146	*One especially provocative*: Colin Williams and Martin Weinberg, "Zoophilia in Men: A Study of Sexual Interest in Animals," *Archives of Sexual Behavior* 32, no. 6 (2004): 523–35.

147	*In Maurice Temerlin's book*: Maurice Temerlin, *Lucy: Growing Up Human* (Palo Alto, Calif.: Science and Behavior Books, 1975).

Asexuals Among Us

150	*"I would say I've"*: Nicole Prause and Cynthia A. Graham, "Asexuality: Classification and Characterization," *Archives of Sexual Behavior* 36, no. 3 (2007): 344.

150	*"I just don't feel"*: Kristin S. Scherrer, "Coming to an Asexual Identity: Negotiating Identity, Negotiating Desire," *Sexualities* 11, no. 5 (2008): 626.

151	*In 2004, Bogaert*: Anthony F. Bogaert, "Asexuality: Prevalence and Associated Factors in a National Probability Sample," *Journal of Sex Research* 41, no. 3 (2004): 279–87.

152	*"They were not particularly"*: Prause and Graham, "Asexuality": 344.

Foot Play: Podophilia for Prudes

156	*"In a small but not inconsiderable"*: Havelock Ellis, *Studies in the Psychology of Sex* (online-ebooks.info, 2004), 5: 12.

157	*reports on male homosexual foot fetishism*: Martin S. Weinberg, Colin J. Williams, and Cassandra Calhan, "Homosexual Foot Fetishism," *Archives of Sexual Behavior* 23, no. 6 (1994): 611–26.

158	*In a subsequent article*: Martin S. Weinberg, Colin J. Williams, and Cassandra Calhan, " 'If the Shoe

Fits . . .': Exploring Male Homosexual Foot Fetishism," *Journal of Sex Research* 32, no. 1 (1995): 17–27.

159 *Ellis admonishes us*: Ellis, *Studies in the Psychology of Sex*, 5:19.

159 *One especially vivid example*: Jules R. Bemporad, H. Donald Dunton, and Frieda H. Spady, "The Treatment of a Child Foot Fetishist," *American Journal of Psychotherapy* 30, no. 2 (1976): 303–16.

161 *About a decade later*: Juliet Hopkins, "A Case of Foot and Shoe Fetishism in a 6-Year-Old Girl," in *The Borderline Psychiatric Child: A Selective Integration*, ed. Trevor Lubbe (London: Routledge, 2000), 109–29.

162 *This live-and-let-live approach*: Joseph R. Cautela, "Behavioral Analysis of a Fetish: First Interview," *Journal of Behavioral and Experimental Psychiatry* 17, no. 3 (1986): 161–65.

163 *This is the intriguing*: A. James Giannini et al., "Sexualization of the Female Foot as a Response to Sexually Transmitted Epidemics: A Preliminary Study," *Psychological Reports* 83, no. 2 (1998): 491–98.

A Rubber Lover's Tale

167 *"When I was four"*: Narcyz Lukianowicz, "Imaginary Sexual Partner: Visual Masturbatory Fantasies," *Archives of General Psychiatry* 3, no. 4 (1960): 432.

168 *As reported in their*: Thomas J. Fillion and Elliott M. Blass, "Infantile Experience with Suckling Odors Determines Adult Sexual Behavior in Male Rats," *Science* 231, no. 4739 (1986): 729–31.

Female Ejaculation: A Scientific Road Less Traveled

173 *In an extraordinary 2010 review*: Joanna B. Korda, Sue W. Goldstein, and Frank Sommer, "The History of Female Ejaculation," *Journal of Sexual Medicine* 7, no. 5 (2010): 1965–75.

176 *report "copious" amounts*: Amy L. Gilliland, "Women's Experiences of Female Ejaculation," *Sexuality and Culture* 13, no. 3 (2009): 121–34.

177 *a team of Egyptian researchers*: Ahmed Shafik et al., "An Electrophysiologic Study of Female Ejaculation," *Journal of Sex and Marital Therapy* 35, no. 5 (2009): 337–46.

177 *a team of Czechs*: Milan Zaviačič et al., "Female Urethral Expulsions Evoked by Local Digital Stimulation of the G-Spot: Differences in the Response Patterns," *Journal of Sex Research* 24, no. 1 (1988): 311–18.

177 *dubious 1966 assertion*: William H. Masters and Virginia E. Johnson, *Human Sexual Response* (New York: Little, Brown, 1966).

178 *"Before he'd say"*: Gilliland, "Women's Experiences of Female Ejaculation," 126.

Studying the Elusive "Fag Hag": Women Who Like Men Who Like Men

179 *her Wikipedia entry*: Rue McClanahan, en.wikipedia.org/wiki/Rue_McClanahan (accessed June 14, 2011).

180 *Nancy Bartlett and her colleagues*: Nancy H. Bartlett et al., "The Relation Between Women's Body Esteem and Friendships with Gay Men," *Body Image* 6, no. 3 (2009): 235– 41.

181 *"The gay man in your life"*: Margaret Cho, *I'm the One That I Want* (New York: Ballantine Books, 2002), 37.

Darwin's Mystery Theater Presents . . . The Case of the Female Orgasm

186 *"Female orgasm is a variable"*: Cindy M. Meston et al., "Women's Orgasm," *Annual Review of Sex Research* 15 (2004): 174.

186 *Gould fleshed out*: Stephen Jay Gould, "Male Nipples and Clitoral Ripples," in *Bully for Brontosaurus*:

Further Reflections in Natural History (New York: W. W. Norton, 1992), 124–38.

187 *it was Lloyd who*: Elisabeth A. Lloyd, *The Case of the Female Orgasm* (Cambridge, Mass.: Harvard University Press, 2005).

187 *Twin-based evidence*: Kate M. Dunn, Lynn F. Cherkas, and Tim D. Spector, "Genetic Influences on Variation in Female Orgasmic Function: A Twin Study," *Biology Letters* 1, no. 3 (2005): 260–63.

188 *"just because something"*: David P. Barash, "Let a Thousand Orgasms Bloom! A Review of *The Case of the Female Orgasm* by Elisabeth A. Lloyd," *Evolutionary Psychology* 3 (2005): 351.

188 *Religiosity is another*: Sheryl A. Kingsberg and Jeffrey W. Janata, "Female Sexual Disorders: Assessment, Diagnosis, and Treatment," *Urologic Clinics of North America* 34, no. 4 (2007): 497–506.

188 *Using self-report data*: Todd K. Shackelford et al., "Female Coital Orgasm and Male Attractiveness," *Human Nature* 11, no. 3 (2000): 299–306.

188 *orgasm and the physical attractiveness*: Randy Thornhill et al., "Human Female Orgasm and Mate Fluctuating Asymmetry," *Animal Behaviour* 50, no. 6 (1995): 1601–15.

189 *"During the female copulatory"*: Danielle Cohen and Jay Belsky, "Avoidant Romantic Attachment and Female Orgasm: Testing an Emotion-Regulation Hypothesis," *Attachment and Human Development* 10, no. 1 (2008): 1.

189 *Chinese women who were dating*: Thomas Pollet and Daniel Nettle, "Partner Wealth Predicts Self-Reported Orgasm Frequency in a Sample of Chinese Women," *Evolution and Human Behavior* 30, no. 2 (2009): 146–51.

190 *"is a signal whereby"*: Barash, "Let a Thousand Orgasms Bloom!," 349.

The Bitch Evolved: Why Are Girls So Cruel to Each Other?

193 *"Jo is a fifteen-year-old girl"*: Rosalyn Shute, Laurence Owens, and Phillip Slee, " 'You Just Stare at Them and Give Them Daggers': Nonverbal Expressions of Social Aggression in Teenage Girls," *International Journal of Adolescence* 10, no. 4 (2002): 353–72.

194 *which I'll summarize here*: Nicole H. Hess and Edward H. Hagen, "Sex Differences in Indirect Aggression: Psychological Evidence from Young Adults," *Evolution and Human Behavior* 27 (2006): 231– 45.

Never Ask a Gay Man for Directions

203 *In a study reported*: Qazi Rahman, Davinia Andersson, and Ernest Govier, "A Specific Sexual Orientation–Related Difference in Navigation Strategy," *Behavioral Neuroscience* 119, no. 1 (2005): 311–16.

203 *in a follow-up study*: Qazi Rahman and Johanna Koerting, "Sexual Orientation–Related Differences in Allocentric Spatial Memory Tasks," *Hippocampus* 18, no. 1 (2008): 55– 63.

204 *different armpit odors*: Ivanka Savic et al., "Smelling of Odorous Sex Hormone–Like Compounds Causes Sex-Differentiated Hypothalamic Activations in Humans," *Neuron* 31, no. 4 (2001): 661– 68.

"Single, Angry, Straight Male . . . Seeks Same": Homophobia as Repressed Desire

205–6 *One of the most important*: Henry E. Adams, Lester W. Wright Jr., and Bethany A. Lohr, "Is Homophobia Associated with Homosexual Arousal?," *Journal of Abnormal Psychology* 105, no. 3 (1996): 440–45.

206 *"a mercury-in-rubber . . . gauge"*: Ibid., 441.

208 *"We believe it is inaccurate"*: Brian P. Meier et al., "A Secret Attraction or Defensive Loathing? Homophobia, Defense, and Implicit Cognition," *Journal of Research*

in Personality 40, no. 4 (2006): 388.

209 *Some of the most startling data*: Gregory M. Herek, *Stigma and Sexual Orientation: Understanding Prejudice Against Lesbians, Gay Men, and Bisexuals* (Thousand Oaks, Calif.: Sage, 1998).

209 *a later study published*: Jeffrey A. Bernat et al., "Homophobia and Physical Aggression Toward Homosexual and Heterosexual Individuals," *Journal of Abnormal Psychology* 110, no. 1 (2001): 179–87.

Baby-Mama Drama-less Sex: How Gay Heartbreak Rains on the Polyamory Parade

215 *"abandoned lovers are"*: Helen E. Fisher, "Broken Hearts: The Nature and Risks of Romantic Rejection," in *Romance and Sex in Adolescence and Emerging Adulthood: Risks and Opportunities*, ed. Ann C. Crouter and Alan Booth (Mahwah, N.J.: Lawrence Erlbaum, 2006), 13.

218 *"a same-sex infidelity"*: Brad J. Sagarin et al., "Sex Differences (and Similarities) in Jealousy: The Moderating Influence of Infidelity Experience and Sexual Orientation of the Infidelity," *Evolution and Human Behavior* 24, no. 1 (2003): 18.

219 *"When asked in a 2010 interview"*: Boris Kachka, "The Kid Stays in the Picture," *New York*, May 16, 2010, nymag.com/arts/theater/features/66008/.

Top Scientists Get to the Bottom of Gay Male Sex Role Preferences

222 *Several years ago*: Trevor A. Hart et al., "Sexual Behavior Among HIVPositive Men Who Have Sex with Men: What's in a Label?," *Journal of Sex Research* 40, no. 2 (2003): 179–88.

223 *a study showing that tops*: David A. Moskowitz, Gerulf Rieger, and Michael E. Roloff, "Tops, Bottoms, and Versatiles," *Sexual and Relationship Therapy* 23, no. 3

(2008): 191–202.

223 *"Although self-labels"*: Hart et al., "Sexual Behavior Among HIV-Positive Men Who Have Sex with Men": 188.

224 *"such relationships also"*: Moskowitz, Rieger, and Roloff, "Tops, Bottoms, and Versatiles": 199.

224 *Another intriguing study*: Matthew H. McIntyre, "Letter to the Editor: Digit Ratios, Childhood Gender Role Behavior, and Erotic Role Preferences of Gay Men," *Archives of Sexual Behavior* 32, no. 6 (2003): 495–97.

Is Your Child a "Pre-homosexual"? Forecasting Adult Sexual Orientation

227 *"was to review the evidence"*: J. Michael Bailey and Kenneth J. Zucker, "Childhood Sex-Typed Behavior and Sexual Orientation: A Conceptual Analysis and Quantitative Review," *Developmental Psychology* 31, no. 1 (1995): 44.

227 *list of innate sex differences*: Ibid.

230 *interviewed twenty-five adult women*: Kelley D. Drummond et al., "A Follow-Up Study of Girls with Gender Identity Disorder," *Developmental Psychology* 44, no. 1 (2008): 34–45.

231 *"those targets who"*: Gerulf Rieger et al., "Sexual Orientation and Childhood Gender Nonconformity: Evidence from Home Videos," *Developmental Psychology* 44, no. 1 (2008): 53.

232 *Cross-cultural data show*: Fernando Luiz Cardoso, "Recalled Sex-Typed Behavior in Childhood and Sports' Preferences in Adulthood of Heterosexual, Bisexual, and Homosexual Men from Brazil, Turkey, and Thailand," *Archives of Sexual Behavior* 38, no. 5 (2008): 726–36.

232 *in a rather stunning case*: Helen W. Wilson and Cathy Spatz Wisdom, "Does Physical Abuse, Sexual Abuse, or

Neglect in Childhood Increase the Likelihood of Same-Sex Sexual Relationships and Cohabitation? A Prospective 30-Year Follow-Up," *Archives of Sexual Behavior* 39, no. 1 (2010): 63–74.

Good Christians (but Only on Sundays)

239 *In my book The Belief Instinct*: Jesse Bering, *The Belief Instinct: The Psychology of Souls, Destiny, and the Meaning of Life* (New York: W. W. Norton, 2011).

240 *"If supernatural punishment is held"*: Dominic Johnson and Jesse Bering, "Hand of God, Mind of Man: Punishment and Cognition in the Evolution of Cooperation," *Evolutionary Psychology* 4 (2006): 219–33.

241 *This is a term coined*: Azim F. Shariff and Ara Norenzayan, "God Is Watching You: Priming God Concepts Increases Prosocial Behavior in an Anonymous Economic Game," *Psychological Science* 18, no. 9 (2007): 803–9.

241 *More recent work*: Will Gervais and Ara Norenzayan, "Like a Camera in the Sky? Thinking About God Increases Public Self-awareness and Socially Desirable Responding," *Journal of Experimental Social Psychology* (in press).

243 *"This approach helps to shift"*: Deepak Malhotra, "(When) Are Religious People Nicer? Religious Salience and the 'Sunday Effect' on Pro-Social Behavior," *Judgment and Decision Making* 5, no. 2 (2010): 139.

244 *In crunching the salacious*: Benjamin Edelman, "Red Light States: Who Buys Online Adult Entertainment?," *Journal of Economic Perspectives* 23, no. 1 (2009): 209–20.

God's Little Rabbits: Believers Outreproduce Nonbelievers by a Landslide

249 *"In the long run"*: Michael Blume, "The Reproductive

Benefits of Religious Affiliation," in *The Biological Evolution of Religious Mind and Behaviour*, ed. E. Voland and W. Schiefenhövel (Berlin: Springer Frontiers Collection, 2009), 122.

252 *"The results are"*: Ibid., 119.

253 *"Some naturalists are trying"*: Ibid., 125.

Planting Roots with My Dead Mother

256 *Consider that before*: "Natural burial," en.wikipedia.org/wiki/Natural_burial (accessed June 14, 2011).

257 *"You could drive about"*: www.naturallegacies.org (accessed June 14, 2011).

257 *This was a term*: Ernest Becker, The Denial of Death (New York: Free Press, 1973).

Being Suicidal: Is Killing Yourself Adaptive? That Depends: Suicide for Your Genes' Sake (Part I)

266 de*Catanzaro posited that human brains*: Denys deCatanzaro, *Suicide/SelfDamage Behavior*, Studies in Archaeological Science (New York: Academic Press, 1981).

267 *But when biologists looked*: Maydianne C. B. Andrade, "Sexual Selection for Male Sacrifice in the Australian Redback Spider," *Science 5*, no. 5245 (1996): 70–72.

268 *Another example is bumblebees*: Robert Poulin, "Altered Behaviour in Parasitized Bumblebees: Parasite Manipulation or Adaptive Suicide?," *Animal Behaviour* 44, no. 1 (1992): 176.

268 *To crystallize his position*: Denys deCatanzaro, "A Mathematical Model of Evolutionary Pressures Regulating Self-Preservation and Self-Destruction," *Suicide and Life-Threatening Behavior* 16, no. 2 (1986): 166–81.

269 *In a 1995 study*: Denys deCatanzaro, "Reproductive Status, Family Interactions, and Suicidal Ideation:

Surveys of the General Public and High-Risk Groups,"
Ethology and Sociobiology 16, no. 5 (1995): 385–94.

270 *"the observational nature of"*: Ibid., 391.

270 *A few years after*: R. Michael Brown et al., "Evaluation of an Evolutionary Model of Self-Preservation and Self-Destruction," *Suicide and LifeThreatening Behavior* 29, no. 1 (1999): 58–71.

271 *"[She] was described as being"*: Kimberly A. Van Orden et al., "The Interpersonal Theory of Suicide," *Psychological Review* 117, no. 2 (2010): 585.

273 *"The adoption of a more dangerous"*: Poulin, "Altered Behaviour in Parasitized Bumblebees: Parasite Manipulation or Adaptive Suicide?

Being Suicidal: What It Feels Like to Want to Kill Yourself (Part II)

275 *According to the researchers*: David Cohen and Angèle Consoli, "Production of Supernatural Beliefs During Cotard's Syndrome, a Rare Psychotic Depression," *Behavioral and Brain Sciences* 29, no. 5 (2006): 468–70.

275 *Some recent evidence suggests*: Anders Helldén et al., "Death Delusion," *British Medical Journal* 335, no. 7633 (2007): 1305.

275 *"The delusion consisted of"*: Cohen and Consoli, "Production of Supernatural Beliefs During Cotard's Syndrome, a Rare Psychotic Depression," 469.

276 *I don't think any scholar*: Roy F. Baumeister, "Suicide as Escape from Self," *Psychological Review* 97, no. 1 (1990): 90–113.

278 *"A large body of evidence"*: Ibid., 95.

278 *"it is apparently the size"*: Ibid., 95.

280 *"The essence of self-awareness"*: Ibid., 98.

280 *"Our best route to understanding"*: Edwin S. Shneidman, *The Suicidal Mind* (New York: Oxford University Press, 1996), 6.

281 *"his file contained"*: Susanne Langer, Jonathan
     Scourfield, and Ben Fincham, "Documenting the Quick
     and the Dead: A Study of Suicide Case Files in a
     Coroner's Office," *Sociological Review* 56, no. 2
     (2008): 304.
281 *"Concluding simply that depression"*: Baumeister,
     "Suicide as Escape from Self," 90.
283 *"Thus suicidal people resemble"*: Ibid., 100.
284 *"When preparing for suicide"*: Ibid., 108.
285 *"acquired capability for suicide"*: Kimberly A. Van
     Orden et al., "The Interpersonal Theory of Suicide,"
     Psychological Review 117, no. 2 (2010): 585.

"Scientists Say Free Will Probably Doesn't Exist, Urge 'Don't Stop Believing!' "

288 *"At the core of"*: Roy F. Baumeister, "Free Will in
     Scientific Psychology," *Perspectives on Psychological
     Science* 3, no. 1 (2008): 14.
290 *The first study to directly*: Kathleen D. Vohs and
     Jonathan W. Schooler, "The Value of Believing in Free
     Will," *Psychological Science* 19, no. 1 (2008): 49–54.
290 *"'You,' your joys and"*: Francis Crick, *The Astonishing
     Hypothesis: The Scientific Search for the Soul* (New
     York: Scribner, 1994), 3.
291 *"If exposure to deterministic"*: Vohs and Schooler,
     "Value of Believing in Free Will": 54.
294 *"Some philosophical analyses"*: Roy F. Baumeister, E. J.
     Masicampo, and C. Nathan DeWall, "Prosocial Benefits
     of Feeling Free: Disbelief in Free Will Increases
     Aggression and Reduces Helpfulness," *Personality and
     Social Psychology Bulletin* 35, no. 2 (2009): 267.

The Rat That Wouldn't Stop Laughing: Joy and Mirth in the Animal Kingdom

298 *"the possibility that our"*: Jaak Panksepp,
     "Neuroevolutionary Sources of Laughter and Social

Joy: Modeling Primal Human Laughter in Laboratory Rats," *Behavioural Brain Research* 182, no. 2 (2007): 232.

299 *"Having just concluded"*: Ibid., 235.

299 *"one must be adept"*: Ibid., 234.

300 *"unless they have been tickled"*: Ibid., 235.

301 *"We have tried to"*: Ibid., 235.

301 *"If a cat"*: Ibid., 241.

302 *In the same issue*: Martin Meyer et al., "How the Brain Laughs: Comparative Evidence from Behavioral, Electrophysiological, and Neuroimaging Studies in Human and Monkey," *Behavioural Brain Research* 182, no. 2 (2007): 245– 60.

303 *the first experimental evidence*: Diana P. Szameitat et al., "Differentiation of Emotions in Laughter at the Behavioral Level," *Emotion* 9, no. 3 (2009): 397– 405.

303 *"the actors were instructed"*: Ibid., 398.

305 *"Schadenfreude laughter might"*: Ibid., 403.

Acknowledgments

Many people poked me in the ribs in response to things I've written in this book. Much to his chagrin, my partner, Juan Quiles, makes an appearance from time to time, and I'm very grateful to him for serving as a muse, a critic, and, more generally speaking, the ever-mysterious yin to my yang. He's one of the few people who have managed to keep me continuously guessing (which means that he brings a healthy chaos that I'm ever in need of).

My agent, Peter Tallack of the Science Factory, has been a fantastic ally working tirelessly behind the scenes. I'm very fortunate to have him representing me, not only because I think he's one of the best agents working the science world today—that makes him sound like a pimp; my profound apologies, Peter— but also because he usually agrees with me. That's what he leads me to believe, anyway.

I'm also lucky to have collaborated on this project with a wonderful team of editors and proofreaders. Most notably, my lead editor, Amanda Moon of Farrar, Straus and Giroux, and her wonderful assistant, Karen Maine, have been at the helm in organizing this collection. Amanda represents *Person Number One* in the editorial process; as my first-pass reader, she's the person who evaluates the strengths and, certainly, the many weaknesses of my essays before anyone else can point them out to me. I feel as if I should include one of those disclaimers about "the attitudes and opinions here are the author's alone and do not necessarily reflect those of his employer." But you know what I mean. We're both, you and I, in very good hands with her.

Another important set of editors is the one that came long before this book ever materialized, when earlier versions of many of the essays were published online. In going to bat for me in conceptualizing and implementing my Bering in Mind column at *Scientific American*, Karen Schrock is the one who really got me off the ground. I can't thank her enough for giving me the outlet to exercise my lewd and lascivious thoughts—through *Scientific American*, no less.

More recently, my editor at *Slate* magazine, Daniel Engber, has also been instrumental in helping me converse with readers about so many gloriously inappropriate topics. Dan and I share a penchant for the absurd and the scientific, two things that go together so naturally and can be such a jubilant concoction when things go just right. I'm enormously grateful for working alongside Dan and learning from him, in his role as an editor but also as a fellow writer.

What would I be without my family? Much worse for the wear, certainly. My guess is that for a very long time, my family members haven't known quite how to answer the question of what it is that I do for a living. Thank you regardless, Dad, Linda, Stacey(s), Adam, Jodi, Jakob, Gianni, Sydney, and those among the many close but scattered strands of Berings and Roths.

Perhaps the most important individuals to thank, however, are the scientists and scholars who did the actual hard work. My summaries herein are only pale accounts of their ingenuity and, often, their genius. Among those I'd like to mention specifically for their contributions and help along the way: Gordon Gallup, Becky Burch, Ray Blanchard, Ara Norenzayan, Denys deCatanzaro, Roy Baumeister, and Michael Blume. Jonathan Jong was a resource-laden assistant who kindly zoomed more than one obscure article to me via the New Zealand ether. I'd also like to thank the staff at the Kinsey Institute in Bloomington, Indiana, for hosting me as an independent scholar.

Finally, allow me to reflect in closing, in sincere and smiling gratitude, upon some heroes of my everyday life, my best nonhuman friends, "Big Tommy," Gulliver, and Uma.

Index

PERV
The sexual deviant in all of us
Jesse Bering

Perv is an evolutionarily informed psychological analysis of humanity's weird and wonderfully tense relationship with its own sexuality. Humans have grappled with troubling aspects of their wanton lust from the earliest recorded times. Jesse Bering here goes where few scientists have gone before and confronts the most taboo issues of human sexuality head on.

Why is the concept of sexual deviance so easily conflated with immorality and disgust?

Is perversion culturally relative or are the same behaviours regarded as wrong in all societies?

What are the developmental origins of sexual deviations and perversions?

Are sexual deviants born or made?

Fascinating, controversial and yet deeply entertaining, *Perv* is the kind of book that you will never forget reading.

FEBRUARY 2014